Decoherence and Quantum Measurements

Decoherence and Quantum Measurements

Mikio Namiki
Department of Physics, Waseda University, Tokyo, Japan

Saverio Pascazio
Dipartimento di Fisica, Università di Bari, Italy

Hiromichi Nakazato
Department of Physics, Waseda University, Tokyo, Japan

World Scientific
Singapore • New Jersey • London • Hong Kong

Published by

World Scientific Publishing Co. Pte. Ltd.

5 Toh Tuck Link, Singapore 596224

USA office: 27 Warren Street, Suite 401-402, Hackensack, NJ 07601

UK office: 57 Shelton Street, Covent Garden, London WC2H 9HE

Library of Congress Cataloging-in-Publication Data
Namiki, Mikio, 1925–
    Decoherence and quantum measurements / Mikio Namiki, Saverio Pascazio,
Hiromichi Nakazato.
    p. cm.
    Includes bibliographical references and index.
    ISBN-13 978-981-02-3077-7
    ISBN-10 981-02-3077-X
    1. Quanrum theory. 2. Physical measurements. I. Pascazio, Saverio.
II. Nakazato, Hiromichi. III. Title.
QC174.12.N36 1997
530.12--dc21   97-41555
                                                               CIP

British Library Cataloguing-in-Publication Data
A catalogue record for this book is available from the British Library.

Preface

Quantum mechanics is undoubtedly the greatest achievement of physics in the 20th century. We are now enjoying its success in large areas of modern physics and technology: It allows us to comprehend particles, to view the structure of condensed matter and contemplate exciting modern techniques. However, in spite of such success, since its very birth quantum mechanics has undergone many hard controversies over its basic concepts. One of the central cores of such controversies obviously lies in the so-called quantum mechanical "measurement problem," which consists in understanding which kind of physical processes take place in a quantum measurement. This is a book on the quantum measurement problem, mainly addressed to graduate students and to those physicists who have no specific knowledge on this subject. On the other hand, we hope that this book can provide matter for thoughts also to the experts!

Many physicists think of the quantum measurement process according to a simple projection rule, which is widely used and commonly known as the "postulate of observation." The transition of a quantum state, as implied by the projection rule, is often called *wave function collapse* by measurement. In this book we shall refer to it as *naive* wave function collapse. As will be explained, however, such a naive wave function collapse is unacceptable, on both theoretical and experimental grounds: Therefore, we shall regard as a satisfactory basis for the occurrence of a quantum measurement a *dephasing* process among the branch waves. It goes without saying that a measuring apparatus, which can be used in a quantum mechanical measurement, should be able to perform two functions: provoke dephasing and produce signals (to display the result of the measurement). The wave function collapse by measurement is completed by dephasing. The latter function is merely a secondary process triggered by the wave function collapse. In this context, the two processes must strictly be distinguished: Even though the latter is very important from the technical point of view, we contend that the wave function collapse by measurement is essentially given by dephasing. For this reason, in this book, we shall mainly discuss the wave function collapse as a decoherence or dephasing process: The main purpose of our investigation will be to disclose and elucidate the role played by a detector as a dephaser, in a quantum measurement process. Also, the expressions "decoherence" and "dephasing" will be used interchangeably. We shall seek an answer to the following question: Which kind of decoherence or dephasing process will yield the wave function collapse by measurement?

Immediately after the birth of quantum mechanics, many eminent physicists were very excited by the quantum measurement problem, because it seemed to disclose a new horizon for epistemology in the microscopic world. Sometimes, the main interest on this problem was shifted to philosophy rather than physics. This trend provoked a sort of repulsion for many physicists, who just did not want to think about the measurement problem. In addition to this, the booming development of quantum mechanics in various fields has left the measurement problem unsolved (or, to say the least, unclear) for quite a long time. During the last two decades, however, the remarkable innovation of experimental technique has enabled us to perform experiments which could previously be considered only at the *gedanken* level. The technological and experimental progress is really "pulling" the measurement problem back to physics, in connection to very concrete topics. One of our purposes, in writing this book, is to give a sound theoretical basis, which enables us to deal with the physical processes taking place in a quantum measurement.

About 20 years ago, S. Machida and one of the present authors (M.N.) proposed the "many-Hilbert-space theory" of quantum measurements, which provides a reasonable scheme for discussing quantum measurements. Throughout this book, we would like to lay down, starting from this theory, a firm theoretical basis, which should enable us to solve the most important practical problems. An introductory description and a critical review of other theories are also included, in order to give a general survey and a ground for comparison. However, this book reflects our personal point of view on the quantum measurement problem and has no pretension of being exhaustive: We did not even attempt to provide a complete overview of all the existing theories and did not mention many other interesting approaches to this issue. Any advice and criticisms, in particular from alternative points of view, are most welcome.

This work was partially supported by the Japanese Ministry of Education, Science and Culture, under the grants: Grant-in-Aid for Scientific Research of the Ministry of Education, Science and Culture (No. 04244105) and Monbusho International Scientific Research Program: Joint Research (No. 08044097), by the Japan Society for the Promotion of Science, by the Italian Institute for Nuclear Physics and by the Italian Research National Council. We also acknowledge the support of Waseda University, Waseda University Grant for Special Research Projects No. 96A-126, of Bari University and of the Commission of the European Communities during the initial period of our joint collaboration. We are grateful to Dr. Paul Haines for helpful advice and a careful reading of the manuscript, to Dr. A. Tonomura for the photographs of his electron beam experiment and to Iwanami Shoten Publishers for their help in preparing some of the figures.

May 1997

Mikio Namiki
Saverio Pascazio
Hiromichi Nakazato

Contents

Chapter 1

GENERAL AND HISTORICAL SURVEY

Let us first describe some general characteristics common to both classical and quantum mechanical measurement processes. We shall then give a brief historical account and discuss some future prospects in the quantum measurement theory. Throughout this book, we shall discuss the quantum measurement problem mainly from the point of view of the many-Hilbert-space theory, which was originally proposed by S. Machida and one of the present authors (M.N.) and has later been developed by many authors. In this chapter, we would like to describe the main flow of thought, without resorting to technical details, which will be supplied in the subsequent chapters.

For a good review on the quantum measurement problem see Ref. 204; Two interesting books on this subject are Refs. 34 and 32. For the many-Hilbert-space theory see, for instance, Refs. 117, 142, 146 for the general ideas and Ref. 144 for a review.

1   General remarks

Most classical and quantum mechanical measurement processes can be decomposed into a set of yes-no experiments, each of which consists of two successive steps, the *spectral decomposition* followed by the *detection*. Even though there are some exceptional cases of measurement processes in which the above two steps are not so clearly separated, we should notice that both functions are essential in any measurement apparatus. For this reason, without a serious lack of generality, we shall assume throughout this book that every measurement process is decomposed into the above-mentioned two successive steps. We shall ask which kind of physical processes we can call a "measurement."

1.1   Classical measurements

We first examine classical measurements from the above-mentioned point of view. As an elementary example, let us consider the measurement of the length of a macroscopic body. Suppose that its length is roughly expected to be 4 cm. In this case, we first observe its length with a scale, for example, to check whether it is longer or shorter than 4 cm. The two results (longer or shorter than 4 cm) are mutually exclusive events. Before the measurement, each of them is to be regarded as an exclusive probabilistic event with probability $P_>$ or $P_<$ (where $P_> + P_< = 1$). In the former case, we proceed to the next measurement, and observe whether it is longer or shorter than 4.1 cm. In the former case, we repeat the procedure and finally reach a final value, say 4.123 cm. We stress that we should regard each measurement as a probabilistic event before actually performing the measurement, even though any classical system obeys deterministic dynamics.

This kind of idea is easily applied to the position measurement of a classical point particle in a spatial region, say Ω. We first divide the region into two parts, Ω_1 and Ω_2, and then ask whether the particle is in Ω_1 or Ω_2. If we find it in Ω_1, we proceed to the next measurement in order to observe it in one of its subregions, say Ω_{11} or Ω_{12}. We can repeat this kind of procedure and reach a precise location of the classical particle within a desired accuracy.

Summarizing, we first divide Ω into (nonoverlapping) subregions, say I_1, I_2, $\cdots$, I_N, and then try to find the particle in one of them. I_i stands for a spatial region around r_i with a certain small volume. See Fig. 1.1. Needless to say, the volume of I_i gives us the accuracy of the position measurement at r_i. In this case, we can regard this division of space as the spectral decomposition step and the act of finding the particle in one subregion as the detection step, respectively. In this type of position measurement, we have decomposed the whole measurement into a series of many successive yes-no experiments.

Remember that the two events of finding the particle in one of two separate regions, say I_i and I_j ($i \neq j$), are mutually exclusive, because one classical particle can never exist at different positions. For this reason, the total probability expected before the measurement is written as a simple sum:

$$P = \sum_k P_k, \tag{1.1}$$

where P_k stands for the probability of finding the particle in the kth subregion I_k. If there is no leakage, we can put $P = 1$.

The same idea can be applied to the momentum measurement of a classical particle. For conventional momentum measurements, we usually deflect the charged particle via a uniform magnet M, as schematically shown in Fig. 1.2. Suppose that a particle with momentum p_i is in an incident beam before entering M and is subsequently deflected into another direction with a definite momentum q_i ($|q_i| = |p_i|$). In this experiment many detectors, D_1, D_2, $\cdots$, D_N, are placed on a sphere, which is remote from the magnet, and the particle is detected by one of them. Detector D_i is located so as to catch the particle

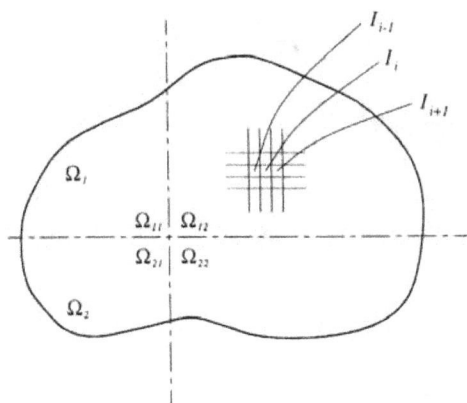

Fig. 1.1. Position measurement.

with momentum in the ith momentum region I_i, around q_i with width $\triangle q_i$ (corresponding to momentum around p_i with width $\triangle p_i$ before entering M). The width of the ith momentum region I_i is given by the aperture of D_i, and determines the accuracy of the momentum measurement at p_i. Needless to say, going through the magnet is nothing but the spectral decomposition step, followed by the detection step at D_i. In this type of momentum measurement, we have decomposed the whole measurement process into a parallel set of many yes-no experiments.

Because one classical point particle cannot simultaneously have two different momenta, we never find the particle in two states with different momenta; the events are mutually exclusive. For this reason, (1.1) still holds if we identify P_i with the probability of finding the particle in the ith momentum region I_i.

The detection step is performed via classical physical processes, such as an observation by the eyes of the observer or by means of an instrument. Even though a classical system obeys deterministic dynamics, Eq. (1.1) clearly displays the probabilistic characteristic that only one of many events will eventually be realized, and also that once one event happens, the other events never happen. Equation (1.1) must be regarded as the theoretical goal of every measurement theory.

There is no serious problem to solve in the case of a classical measurement.

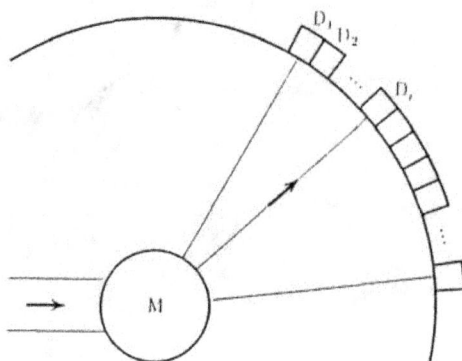

Fig. 1.2. Momentum measurement.

1.2   Quantum mechanical measurements

A quantum mechanical position or momentum measurement is also to be considered along the above-mentioned line of thought, but an important difference arises from the fact that every quantum state is to be described in terms of the Schrödinger wave function. The spectral decomposition step is now written as

$$\psi_0 = \sum_i c_i \chi_i^{(0)} \longrightarrow \psi_I = \sum_i c_i \chi_i, \tag{1.2}$$

where $\chi_i^{(0)}$ and χ_i are the wave functions before and after the spectral decomposition and c_i complex numbers. From now on, we will call each summand of ψ_I a *branch wave* corresponding to the spectral decomposition (1.2). Let us make two examples: For the position measurement, since $\psi_I = \psi_0$ in this particular case, we must set $\chi_i^{(0)} = \chi_i$ and identify χ_i with the position wave functions, which vanish outside I_i. On the other hand, for the momentum measurement, one sets

$$\chi_i^{(0)} = u_i \phi, \quad \chi_i = v_i \phi_i, \tag{1.3}$$

where $u_i = \exp(i p_i \cdot r/\hbar)$ and ϕ are, respectively, the p_i-momentum eigenfunction (apart from a trivial normalization constant) and the position wave packet before entering M, while $v_i = \exp(i q_i \cdot r/\hbar)$ and ϕ_i are, respectively, the q_i-momentum eigenfunction and the position wave packet running toward D_i (corresponding to the ith momentum region I_i after passing through M).

Returning to the general case, according to the probabilistic interpretation of quantum mechanics, the probability density of finding the particle immediately before the detection step is given by

$$P_{\mathrm{I}}(x) = \sum_k |c_k|^2 |\chi_k(x)|^2 + \sum_i \sum_{j \neq i} c_i c_j^* \chi_i(x) \chi_j^*(x). \tag{1.4}$$

This quantity is to be compared with (1.1) in the classical case. As is well known, the quantum mechanical probability differs from the simple sum of the probabilities of finding mutually exclusive events because of the presence of the second term, i.e. the *interference* term. This is one of the most remarkable characteristics of quantum mechanics.[a]

Even in the case of a quantum mechanical measurement, we must have, as a final result, a simple sum of probabilities corresponding to mutually exclusive events, if the measurement is to have a definite meaning; in such a case, we will talk of "perfect measurement." For example, if we find the particle in one region I_i, then we cannot find it in any other region I_j ($j \neq i$), in a position or momentum measurement, because of the indivisibility of a particle. In general, once we have observed the quantum mechanical particle in one region (for some observable), by a perfect measurement, we cannot find it in another one. This is the exact meaning of a perfect measurement. Otherwise, we cannot talk about measured values in a definite way. This means that a perfect measurement must bring a quantum mechanical system from a superposed state, represented by (1.2), into a mixed state representing the final result, in which the probability reads

$$P_{\mathrm{F}}(x) = \text{dephasing}\left(|\psi_{\mathrm{I}}(x)|^2\right) = \sum_k |c_k|^2 |\chi_k(x)|^2. \tag{1.5}$$

By integrating over x one obtains

$$P_{\mathrm{F}} = \int dx \, P_{\mathrm{F}}(x) = \sum_k |c_k|^2. \tag{1.6}$$

This is to be compared to the classical case (1.1). It goes without saying that what we wrote at the end of the preceding subsection is still valid.

The transition from a superposed state, whose probability is given by (1.4), to a state which is characterized by the lack of interference term in its probability (1.5), is sometimes called the "wave function collapse" by measurement. Hereafter we will often use WFC as an abbreviation for wave function collapse.

---

[a]Integrating (1.4) over x and making use of orthogonality $(\chi_j, \chi_i) = \delta_{ji}$, we could erase the second term, i.e. the interference term. As will be shown later in Section *1.2* of Chapter 4, however, the disappearance of interference does not necessarily mean that branch waves are not coherent.

However, we also know that many physicists consider the WFC to be a *sudden transition* of the projection type

$$\psi_{\mathrm{I}} \rightarrow \chi_k, \tag{1.7}$$

or equivalently,

all branch waves other than χ_k *suddenly* disappear, $\qquad$ (1.8)

if we find the particle in the kth region, by a measurement performed at time t. We shall henceforth denote this type of transition by the expression "*naive wave function collapse.*" In this case, if we supplement the naive WFC (1.7) or (1.8) by the additional rule

probability of finding the kth event $= |c_k|^2$, $\qquad$ (1.9)

normalized by $\sum_k |c_k|^2 = 1$, then we can easily understand that the final accumulated distribution of all experimental results under the same initial conditions (i.e. the same ψ_{I}) is expressed by (1.5). In other words, the naive WFC (1.7), with the help of (1.9), is a possible solution of the dephasing process (1.5) in terms of wave functions.

On the other hand, we also know that the naive WFC (1.7) can work well as a computational rule, provided it is supplemented by the additional rule (1.9). In fact, all experimental results can be predicted by theoretical calculations performed with this rule. As far as one is contented with a practical rule, and asks no deeper questions, one can talk about any quantum measurement in terms of the naive WFC together with (1.9). For this reason, the naive WFC is widely accepted as a *measurement postulate* among those physicists who regard measurements only from a pragmatic point of view.

In principle, however, we have to make the following objections to the naive WFC: (i) No measurement can be performed *instantaneously*. Any possible measurement can only be realized through physical processes that take place within a very long time duration on the microscopic time scale. (ii) The naive WFC has no theoretical or experimental justification within quantum mechanics, because it gives a definite result only for one measurement performed on a single system. By contrast, as is well known, quantum mechanics does not predict any definite result for a single experiment; rather, it can only give a definite distribution law for the accumulation of many experimental results performed on many independent dynamical systems, each of which is described by the same wave function. Thus, the naive WFC is neither justified by experiment, nor by theory. (iii) The naive WFC (1.7) does not explicitly describe the probabilistic rule (1.9), which is of vital importance for quantum mechanics. Furthermore, it does not explicitly describe the other important fact that the two events of finding the particle in two separate regions, say

I_i and I_j, must be mutually exclusive. In conclusion, the naive WFC, as formalized by (1.7), is not satisfactory. We have to search for a sound physical mechanism that yields (1.5), rather than (1.7). In this context, we can say that (1.7) is merely a short-cut interpretation towards the "wave function collapse." Later (in Chapter 3), we will reformulate the notion of WFC in a proper way.

As for the second point on whether the naive WFC (1.7) can be justified by experiment or not, a related problem, the so-called quantum Zeno effect, has recently excited our interest. Suppose that a dynamical system is in a given state a (which is in general unstable due to transitions into other states) at the initial instant $t = 0$. If the naive WFC (1.7) were true, the quantum mechanical state, if it were observed in the initial state a at a later time $t(> 0)$, would just go back to the initial state. Therefore, one may expect that frequent repetitions of such an experiment would make the system almost stable. Eventually, in the limit of infinite number of repetitions, we are led to the paradoxical conclusion that the unstable system becomes stable. Admittedly, this limiting situation is only mathematical and is not realistic, in practice. This conclusion is named "quantum Zeno paradox," after the famous paradox given by the ancient Greek philosopher Zeno. See Ref. 125. It is now possible to observe this effect: the system will become more and more stable as the number of repetitions of the experiment increases. This is called "quantum Zeno effect."[153] For this reason, the quantum Zeno effect has a close connection with the naive WFC, and indeed there exists, among many physicists, the widespread opinion that the verification of the quantum Zeno effect provides experimental evidence in favour of the naive WFC. The experimental observation of the quantum Zeno effect in decay phenomena is normally very hard to carry out, but it has recently become possible, in some sense, due to a new experimental proposal.[43] After performing this experiment, the experimental group[84] claimed to have been the first to experimentally observe the quantum Zeno effect, by obtaining the same reduction of the transition probability as expected on the basis of the naive WFC. This seemed to be the first experimental verification of the naive WFC. However, this claim was immediately criticized by some theoretical work[160,83,153,152] that showed that the very same result can be derived through a dynamical transition process, without resorting to any sudden projection, as in the naive WFC (1.7). This simply means that observation of the quantum Zeno effect is not necessarily an experimental evidence in support of the naive WFC. Furthermore, a recent paper[133] showed that the above experimental proposal and theoretical analysis are at variance with the genuine "quantum Zeno effect" as originally formulated in Ref. 125. The quantum Zeno effect will be carefully analyzed in Chapter 8. This effect is very interesting from the measurement-theoretical point of view, because it is closely connected to the essential core of the measurement problem: What is the wave function collapse by measurement?

Returning back to our main topic, we repeat that a perfect measurement on a quantum mechanical system, or the WFC, can be realized when and only when the detector erases the interference term in (1.4); more precisely, the detector must provoke quantum dephasing among the branch waves χ_i in ψ_I of (1.2). Consequently, we have to face the crucial problem of understanding which kind of physical process enables the detector to yield such a quantum dephasing. This is the reason why many physicists have been involved for a long time in the quantum measurement problem, although they have never been bothered by the classical measurement problem. The core of the measurement problem is therefore to elucidate the mechanism at the origin of such a dephasing process, within the theoretical framework of quantum mechanics.

1.3   Yes-no experiment

In order to gain a better understanding of the topics discussed above, it is convenient to discuss an interference experiment of the Young type. See Fig. 1.3. Usually, such an interference experiment is set with the intent of observing the wave-particle dualism of a quantum mechanical particle. However, we wish to postpone the analysis of the wave-particle dualism itself to the next chapter, and simply discuss this experiment as a typical "yes-no experiment." In this section, our main interest is focused on which path (i.e. which of the two slits a and b) a particle Q passes through.

Before starting our discussion, we remark that a foundational experiment on a quantum mechanical particle is usually performed by making use of a sufficiently weak incident beam coming out of a particle emitter E: In this way we can send particles, one by one on the average, into the experimental setup and accumulate the results of many independent experiments performed under the same conditions. We obtain in this way the experimental results that are to be compared with the theoretical predictions made by faithful application of quantum mechanics. Suppose that the beam sends each quantum mechanical particle, represented by the wave function ψ_0, towards a barrier with two small slits a and b, so that two waves ψ_a and ψ_b emerge from the slits into the opposite side of the barrier. If the passage through slit a (b) is identified with one of two exclusive measurement propositions, say A (B), the passage through the barrier can be considered as the corresponding spectral decomposition step of the measurement, described by

$$\psi_0 \;\rightarrow\; \psi_I = \psi_a + \psi_b. \tag{1.10}$$

In general, the probability of finding the particle after it has passed through the double-slit barrier reads

$$P_I = |\psi_a|^2 + |\psi_b|^2 + 2\,\mathrm{Re}\,\psi_a\psi_b^*. \tag{1.11}$$

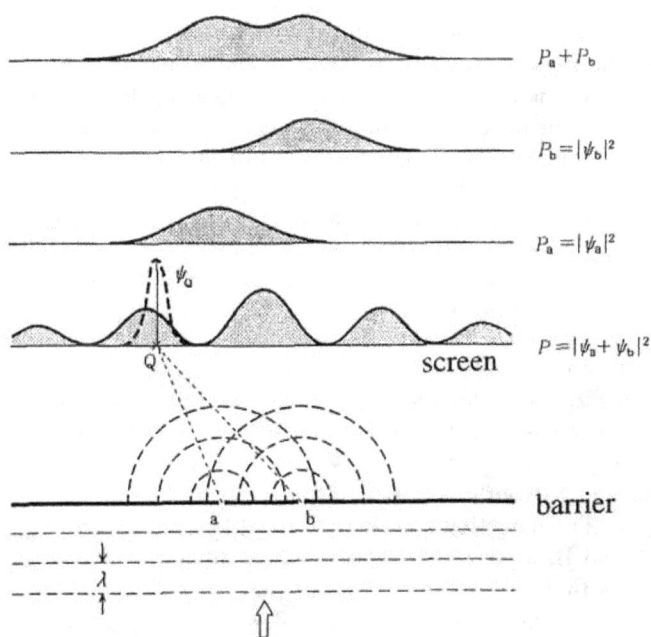

Fig. 1.3. Interference experiment of the Young type.

The detection step is performed by detectors D_a and D_b (not shown in Fig. 1.3), placed after slits a and b, in order to detect the particle path, a or b. If D_a (D_b) gives a signal, then we know that the particle passed through slit a (b). Of course, the two events are mutually exclusive, so that the expected result is represented by the following probability

$$P_F = |\psi_a|^2 + |\psi_b|^2, \qquad (1.12)$$

which is characterized by the lack of the interference term. Therefore, the measurement of the particle path, performed by the detectors on an indivisible particle, must provoke the erasing of the interference term, i.e. the quantum dephasing between the branch waves ψ_a and ψ_b.

This is a typical yes-no experiment which determines the particle path, a or b, i.e. the measurement proposition $\mathcal{A}$ or $\mathcal{B}$. Another good example of yes-no experiment is the Stern-Gerlach experiment to determine a spin state. See Fig. 1.4. By means of a non-uniform magnetic field or a magnetic mirror, the spin wave function of a particle in an indefinite spin state is separated, according to the spin state, into an upper and a lower channels A and B, corresponding to two exclusive measurement propositions $\mathcal{A}(= $ spin up) and $\mathcal{B}(= $ spin down),

$$\psi_0 = (c_a u_a + c_b u_b)\phi \;\rightarrow\; \psi_I = c_a u_a \phi_a + c_b u_b \phi_b. \qquad (1.13)$$

Here, c_a and c_b are c-number constants, u_a and u_b are the up- and down-spin eigenfunctions, ϕ stands for the wave packet before the decomposition and ϕ_a and ϕ_b for the wave packets running towards D_a and D_b, respectively, after the decomposition.

Compare (1.13) with (1.10), under the identification $\psi_a = c_a u_a \phi_a$ and $\psi_b = c_b u_b \phi_b$. The detection step is performed by detectors D_a and D_b put in channels A and B, as in the interference experiment of the Young type. We should remark that in both the above-mentioned cases, we can remove one of the two detectors, say D_b, without any change in the measurement procedure. If D_a generates a signal, i.e. if we perform a coincidence experiment ("yes" both in the emitter E and D_a), we can conclude that the particle passed through channel A, i.e. that the measurement proposition $\mathcal{A}$, corresponding to channel A, is affirmative. On the other hand, in the case of an anti-coincidence experiment ("yes" in E and "no" in D_a), we can conclude that the particle passed through channel B, i.e. that the other measurement proposition $\mathcal{B} = \text{not}\mathcal{A}$, corresponding to channel B, is affirmative. Consequently, we obtain the same conclusions as in the case of a measurement performed when both D_a and D_b are present. The latter case is sometimes called a "negative-result measurement." Remember that in this case we shall often suppress the subscript a of D_a later.

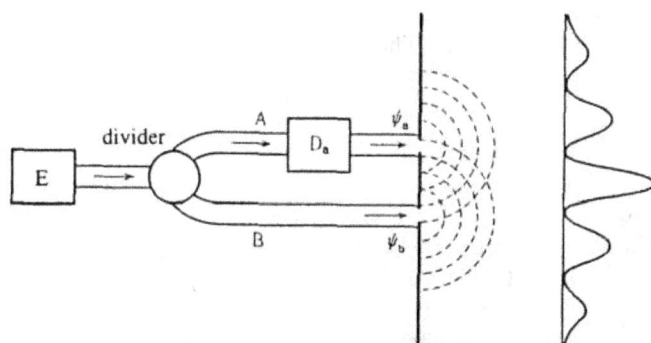

(a) Young type

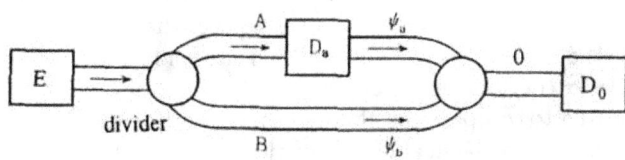

(b) neutron-interferometer type

Fig. 1.4. Stern-Gerlach experiment.

In this type of experiment, we can observe the physical effects provoked by instrument D_a, after recombination of the two branch waves into the final channel 0, by an additional detector D_0 placed in channel 0. In the case of a perfect measurement by D_a, we can expect that the measurement at D_0 does not show any interference. By contrast, if D_a yields no dephasing effect, we observe a beautiful interference pattern at D_0. In the intermediate cases, we should observe at D_0 a partially decohered state between the two branch waves. These intermediate cases will be called "imperfect measurements."

In the case of an interference experiment of the Young type, we can observe this effect by performing a position measurement on the screen (which is to be considered on the same footing as D_0 in the above Stern-Gerlach experiment). Take small spatial regions I_k on the screen and apply the whole argument of Section *1.2* to this situation. The measurement yields the final distribution law

$$|c_k|^2 \rightarrow |\psi(r_k)|^2 d^3 r_k, \tag{1.14}$$

in the limiting case in which the size of I_k becomes very small. We therefore observe a beautiful interference, no interference, or an intermediate phenomenon, if the detector (D_a and/or D_b) provokes no dephasing, perfect dephasing, or partial dephasing (imperfect measurement), respectively. Additional details will be discussed later (in Section *2.2* of Chapter 4).

Throughout the above discussion, we have assumed that all detectors are of the non-destructive type, such as, for example, those invented in order to detect charged particles without destroying them. However, this assumption is too restrictive, because we often use detectors of the destructive type, such as photon or neutron detectors, which generate signals by absorbing (i.e. destroying) incoming photons or neutrons. In such a case, the original particle does not exist after the measurement. We shall discuss and formulate this kind of measurement later (in Section *2.2* of Chapter 4).

It is interesting to compare the above yes-no experiments with coin tossing, as a typical example of a classical probabilistic event. The two sides of the coin are identified with two mutually exclusive measurement propositions, A and B, corresponding to the passage through channel A and B. In principle, we can fully analyze the whole temporal behaviour of the coin by means of a classical and deterministic dynamics, provided that we know every detail of its initial conditions. Nevertheless, we are forced to introduce probabilistic arguments, in order to predict the possible behaviour of the coin, because we do not exactly know its initial conditions. This is a classical probabilistic problem, in which the notion of probability must be used in practice, but not as a matter of principle. By contrast, in general, the notion of probability in quantum mechanics is used in principle and is not simply a practical matter: We cannot pursue the whole temporal behaviour of a quantum mechanical system, in terms of temporal changes of its observables, even if we know every detail of its initial wave function. This is one of the most important differences between classical and quantum systems: Probabilities are epistemic in classical physics and ontologic in quantum mechanics. Nonetheless, both in the classical and quantum case, the measurement process can be described as a probabilistic process and yields

$$P_F = P_A + P_B \tag{1.15}$$

for the probability of obtaining the final result. Remember that the very expression (1.15) signifies that either A or B has occurred, as one of two mutually exclusive events. At the final step, after having tossed the coin, we must obviously observe only one side of the coin. We understand that the probabilistic expression (1.15) naturally includes, as a probabilistic event, even such a final step. The final probabilistic expression for a quantum measurement, such as (1.5) or (1.12), is to be understood in a way analogous to the above-mentioned classical event, even though the two obey a different underlying dynamics.

2 Brief history

In the above discussion, we have stressed that one of the essential parts of a quantum mechanical measurement process is the quantum dephasing among the branch waves. Historically, immediately after the birth of quantum mechanics, Heisenberg[74] and Bohr[27] insisted that the interaction between a microscopic object system Q (subject to quantum mechanics) and a macroscopic measuring apparatus A (described in terms of classical dynamics) is out of human control and provokes enough randomness to yield quantum dephasing among the branch waves. However, we do not find their arguments satisfactory, because even a macroscopic apparatus is composed of many microscopic particles and the latter are fully governed by quantum mechanics.

2.1 Von Neumann-Wigner theory

The von Neumann-Wigner theory of measurement is strongly based on the credo that quantum mechanics (in particular, the superposition principle) must be strictly applied to both the microscopic object Q and the macroscopic apparatus A. These must be treated on an equal footing and along the same line of thought. In this theory, unitarity is not destroyed by any physical measurement process. Consequently, we can never have quantum dephasing, leading to the WFC, by means of a measurement process performed on Q, as will be discussed in detail in Chapter 3. For this reason von Neumann introduced a "chain" of measurements connecting the object to the observer: Interaction of Q with A → observer's act of looking at A → transmission to his nervous system → to his brain cells. ... However, no wave function collapse takes place at any step of the chain, as far as we have only a physical process at each step. Finally, von Neumann and Wigner concluded that the WFC could be provoked by and only by the action of the "Abstraktes Ich" or "consciousness" of the observer (a human being), placed at the end of the chain. We emphasize that in their theory, quantum dephasing, yielding WFC, is rooted in such a metaphysical origin.[195,205]

Schrödinger criticized the von Neumann-Wigner theory by proposing his famous "cat paradox" (Fig. 1.5): Suppose that we have two boxes, the first containing both a small amount of radioactive material and a detector that generates a signal if hit by radiation or by the decay products of the radioactive material, and the second containing a poisonous gas bottle and a cat. Assume that the radioactive material has only two states, the ground state and an excited state. If the radioactive material remains in the excited state, there is no signal from the first box and therefore no poisonous gas in the second box, so that the cat is alive. However, if the transition from the excited state to the ground state takes place, the cat is killed by the gas coming out of the bottle. This means that the excited and ground states correspond exactly to the "alive" and "dead" states of the cat, respectively. Undoubtedly, this instrument is a measuring apparatus that is able to observe the cat state. However, the observer—a human being—can only know the final result (namely whether the cat is alive or dead) by opening a small window and looking at the second box. According to the von Neumann-Wigner theory, the act of looking is nothing but a quantum mechanical measurement, so that the state of the cat (alive or dead) is not definite, before the observer has looked into the box. In other words, the cat state becomes definite, alive or dead, as a consequence of the very act of looking. This situation contradicts our common knowledge that the state of the cat must be definite irrespective of the act of looking. Such a "look" simply displays the information encoded in the measurement, irrespectively of the observer's consciousness. This is usually called Schrödinger's cat paradox.

Wigner's friend paradox is even more interesting than the above paradox, due to Schrödinger. Replace the cat with a human being—a friend of Wigner's. The only essential difference is the fact that the cat, unlike Wigner's friend, has (or is supposed to have) no consciousness. It is better to avoid killing Wigner's friend, by replacing the poisonous gas bottle with an electric lamp. Since the excited and ground states correspond to the "on" and "off" states of the lamp, respectively, this instrument can work as a measuring apparatus for the state of the radioactive material, like in the above Schrödinger's cat example. However, Wigner's friend is inside a black box, so that the final observer must call the friend and ask what the result is (lamp on or off). In this case, a serious question arises in relation to the uniformity of the chain of measurements connecting the object to the observer: When did the WFC take place? Did it take place (i) when the observer called, or (ii) when the

counter

radioactive material   poisonous gas

Fig. 1.5. Schrödinger's cat.

friend looked at the lamp? If the first case were true, the lamp would have to be considered in an indefinite state before the observer calls, and then only the act of calling determines the state of the lamp. The role of the friend is the same as the role of the cat in Schrödinger's paradox. In other words, we face a very strange situation, contradicting our common knowledge that the lamp is in a definite state irrespectively of the call. On the other hand, if the second case were true, the WFC should take place at the very instant the friend realizes the state of the lamp. If so, we are led to the conclusion that the very act of calling should not be included in the theoretical framework of the von Neumann-Wigner theory but is merely a classical extraction of information, so that the uniformity of the chain of measurements is destroyed. We believe that this paradox implies a serious catastrophe of the von Neumann-Wigner theory. In Chapter 4, we shall discuss a possible (experimental) realization of Schrödinger's cat state as a "mesoscopic phenomenon" or an "imperfect measurement."

In the von Neumann-Wigner theory, we have used both a "conscious" human being and a "consciousless" cat. Is the cat really "consciousless"? Even though this question would be very difficult to answer, this problem is a target of life science, which is undergoing a very rapid development. Von Neumann and Wigner could not be aware of this, over 60 years ago. Furthermore, sooner or later brain science will tackle the issue of "consciousness" and/or "free will." Modern science cannot afford to neglect this kind of effort, otherwise physicists could be blamed for their *arrogance* in discarding such sciences.

2.2   *Ergodic amplification theory*

One of the most important antagonists of the von Neumann-Wigner theory was the so-called "ergodic amplification" theory. Usually, the microscopic object systems to be observed are so small that they cannot have enough energy to drive macroscopic measuring apparata or detectors, so that energy must be supplied by some kind of amplification process. In most cases, the amplification process becomes a thermal irreversible process, such as counter discharge or diffusion, and therefore we can expect quantum dephasing leading to WFC.[b] The underlying idea is the rather naive expectation that such a thermal irreversible process can naturally provoke quantum dephasing, and that the conventional derivation of statistical mechanics from quantum mechanics can be formulated by relying upon such quantum dephasing. The idea of the ergodic amplification theory was widely accepted among many physicists, because of its reasonable characteristic that it never introduces metaphysical considerations, as in the von Neumann-Wigner theory, and sharply separates the observer from the object: von Neumann's chain is cut at the measuring apparatus. This measurement theory was also illustrated by means of a simple model[69] and formulated on the basis of the dynamical statistical theory of irreversible processes.[45] Remember that this measurement theory ascribes the essential origin of quantum dephasing, i.e. of the WFC, to the thermal irreversible processes which actually take place inside the measuring apparatus. Some physicists[169] thought that this theory solved the measurement problem.

The ergodic amplification theory was strongly criticized on the basis of (i) Wigner's "No Go" theorem[205] and (ii) the "negative-result measurement" paradox.[90] Wigner's "No Go" theorem states that a quantum system can never reach any mixed state via a unitary evolution, even if the initial state of the measuring apparatus were a mixed state. This theorem countered the rather hasty understanding that quantum dephasing in measurement processes is rooted in the mixed nature of the apparatus at the initial instant. Wigner even suggested that the present formalism of quantum mechanics is unable to yield statistical mechanics.[206] A serious claim!

Some advocates of the ergodic amplification theory made a strong objection and provoked a very interesting debate[170,115] against Wigner and his collaborators. From our viewpoint, however, we should notice that even though Wigner's theorem is mathematically correct, it is only proved within the framework of a *single* Hilbert space. In this book we shall break through the barrier of Wigner's "No Go" theorem and then derive the WFC, by representing the measuring apparatus in terms of a direct sum of *many* Hilbert spaces.

---

[b]Strictly speaking, the WFC (1.5) is the result of *complete* quantum dephasing, while (possibly partial) quantum dephasing does not necessarily imply (1.5). However, unless confusion may arise, we shall use both expressions interchangeably in this book.

The "negative-result measurement" paradox was presented as an argument against the ergodic amplification theory.[168] As already mentioned, the spin-down measurement in a Stern-Gerlach experiment (see Fig. 1.4) is performed without resorting to any thermal irreversible process, such as counter discharge (remember that the detector is placed in the upper channel, corresponding to spin up). This does mean that the WFC is realized irrespectively of actual thermal irreversible processes. It is true that the WFC is a sort of irreversible process, because the measurement process is not reversible, but the very measurement process is to be strictly distinguished from thermal irreversible processes, such as counter discharge. Remember that the latter are only secondary processes taking place in the apparatus, which is designed so as to display the result of the measurement. Note that some important physicists even believed that the ergodic amplification theory was killed by this paradox.[88] For this reason, we have to search for another physical origin of quantum dephasing, leading to the WFC, different from thermal irreversible processes. Note also that most thermal irreversible processes are regarded as physical processes leading to thermal equilibrium; by contrast, the final state, after the WFC, may not be a thermal equilibrium state.

The criticism that the simple model of apparatus in Ref. 69 can never yield the WFC, as pointed out in Ref. 63, is also worth noting. On the other hand, it is possible to conceive a nice example of model detector, in which we need not use any kind of amplification mechanism, relying upon thermal irreversible processes, in order to display the result of the measurement: Suppose that the initial object particle carries an amount of energy which is large enough to drive an apparatus; for example, a particle in a primary cosmic ray is able to produce an extensive air shower with an energy of the order of 10^{19} eV $\sim$ 1J. In this case, we can formulate a simple model of apparatus, yielding the WFC by measurement, without making use of any kind of amplification process.[117,142] This model serves as a counterexample against the ergodic amplification theory.

In conclusion, the ergodic amplification theory is ruled out due to the negative-result measurement paradox: The WFC cannot be ascribed to the thermal irreversible processes taking place inside the detector. Nevertheless, we cannot overlook the fact that the signal, displaying the result of the measurement, plays a very important role. Roughly speaking, a measuring apparatus performing a quantum measurement must have two functions: it must generate a signal and provoke the quantum dephasing. Remember again that the "negative-result measurement" paradox pointed out that the measurement process must be distinguished from any thermal irreversible process, such as counter discharge.

2.3   Other measurement theories and topics

Besides the von Neumann-Wigner theory and the ergodic amplification theory, many measurement theories have so far been proposed. Among them, we should mention the many-worlds interpretation[54,202] and the environment theory.[211] Here we outline the former in some detail, and postpone to Chapter 3 an explanation (not free of criticisms) of the latter.

The many-worlds interpretation starts from the assumption that recording the result of the measurement, for example, the kth result (as in Section 1), drives the whole system, including Q and A, into the kth world, which is split and different from the other worlds corresponding to different values of k. Two branch waves belonging to different worlds cannot interfere with each other, so that the theory considers the splitting of the world itself to be quite enough to yield the wave function collapse. Furthermore, some authors calculated a sort of frequency for the appearance of a certain state if, say, the result is k, and found that the frequency is equal to the quantum mechanical probabilistic prediction, i.e. $|c_k|^2$. On the basis on this result, they claimed to be able to derive from their theory the probabilistic postulate of quantum mechanics.

The above claim is only rhetorical and is not based on any real physical processes. This is the main reason why many physicists have always been skeptical about the many-worlds interpretation. Recently, however, some quantum cosmologists have used this theory by identifying the splitted worlds with the members of the quantum mechanical probability ensemble for the (unique) universe. Application of quantum mechanics to the unique universe is a very hard issue of modern physics, because the "distance" between the microscopic world, for which quantum mechanics is justifiably applied, and the unique universe is huge. Instead of this kind of "short-cut" application of quantum mechanics, we think that one should first solve, step by step, more local cosmological problems by means of general relativity and quantum mechanics. We do not believe that such a serious issue can be simply dealt with by merely reinterpreting quantum mechanics.

As for the "derivation" of the probabilistic postulate by the advocates of the many-worlds interpretation, we are also skeptical for the following reason. In order to claim such a conclusion, one has to introduce the other important assumption that the probability ensemble of quantum mechanics be composed of all members of the splitted worlds. In this context, we simply think that the output is nothing other than the input.

On the other hand, the "environment" theories are based on the orthogonal decomposition of the environment (or apparatus) state and the partial-trace (or partial inner product) technique. Because the explanation and criticism of these theories require the use of technical details, based on mathematical formulae, we postpone our discussion to Chapter 3.

Apart from the measurement theory itself, we have to pay attention to sophisticated new technologies closely connected to the measurement problem. One of these concerns the quantum Zeno effect, outlined in the preceding section and to be discussed in detail in Chapter 8. Another one is the so-called quantum non-demolition measurement, which enables us to perform certain quantum measurements in quantum optics without destroying, in some sense, the initial quantum state. This is accomplished by making use of nonlinear effects. We will discuss these problems later. Besides them, we should also pay attention to mesoscopic physics, a rapidly developing field, and its repercussions on the measurement problem. A detailed discussion on mesoscopic physics, however, would be out of the scope of this book.

2.4   Many-Hilbert-space theory

We have already mentioned that the essential core of the measurement problem is to yield quantum dephasing, leading to the wave function collapse by measurement. Here we outline the basic ideas of the many-Hilbert-space theory, on how to derive quantum dephasing from the fundamental Schrödinger equation applied to the total system composed of Q and D. The main references are Refs. 117, 138, 142, 144, 146.

First of all, we remark that a measuring apparatus or detector (to be denoted by D) is an open macroscopic system composed of many local systems, each of which is still an open macroscopic system. Hereafter, we shall denote one or more of such local systems by the same letter D. The incident microscopic object system (to be denoted by Q) will undergo a complicated interaction with D, which provokes quantum dephasing on Q. On the other hand, we recall that a foundational quantum mechanical measurement is usually carried out by making use of a weak incident beam. In other words, the Q particles come into D, one by one on the average, so that a Q particle, say the $\ell + 1$th one in an experimental run, will enter D some time after the preceding particle, say the ℓth one, has passed through D. There is a time gap between the ℓth and $\ell + 1$th particles; therefore, the ℓth Q particle will meet a local system, say D_ℓ, which is different from $D_{\ell'}$ ($\ell' \neq \ell$), because each local system will change its state due to the microscopic internal motions taking place inside D during the time gap. For this reason, the local system met by the ℓth Q particle has been labeled by the same ℓ. The ℓth local system D_ℓ is distinguished from other local systems labeled by different $\ell' \neq \ell$. For example, the number of constituent particles constituting a local system is different for different ℓ. Consequently, the whole apparatus state cannot be represented by a single Hilbert space, but we have to introduce a direct sum of many Hilbert spaces for the mathematical representation of D. Each member of the collection of many Hilbert spaces contains the quantum mechanical states of the ℓth local system. Therefore, we have to use N_p Hilbert spaces, in order to represent the whole apparatus state, where N_p is the total number of Q particles sent by the incident beam in an experimental run. This is the reason why we call this kind of scheme "many-Hilbert-space" theory.

According to the original idea of quantum mechanics, we have to accumulate the results of many independent experiments, whose statistical distribution is to be compared with the theoretical prediction. In other words, we have to compare the average over ℓ of experimental results with the quantum theoretical distribution. For very large N_p, we can expect that the average over ℓ is replaced with the statistical ensemble average of D. This is a sort of *ergodic* assumption. Usually, we can assume that the total number of constituents of D, say N, is very large.

For a better comprehension, consider a yes-no experiment as in Fig. 1.3 or Fig. 1.4. Due to the above-mentioned characteristics of a quantum mechanical experiment, in general, each Q particle suffers its own phase shift at detector D. Such a phase shift depends on ℓ and is different for other Q particles with different ℓs. When the sequence of these phase shifts becomes random, we can expect to obtain quantum dephasing, leading to a vanishing interference term in (1.11), in the examples of Fig. 1.3 or Fig. 1.4.

When there is no fluctuation in the sequence of phase shifts, it is obvious that we must observe a beautiful interference pattern. When the phase shifts are randomly distributed in the sequence, we may expect to have a perfect measurement, i.e. the WFC. If the degree of randomness in the sequence is intermediate between these two extreme cases, we are faced with *imperfect measurement*. Everything will depend on the degree of randomness of the sequence. For this reason, we can quantitatively discuss the degree of dephasing by means of the so-called *decoherence parameter*, which gives a quantitative estimate of the randomness of the sequence. The decoherence parameter will be introduced in Chapter 4.

As will be discussed later, we expect to observe quantum dephasing, leading to the WFC, in the limit of infinite N (and also of infinite N_p). In other words, the WFC may be regarded as a sort of phase transition or of condensation phenomenon, which can be observed only in the limit of infinite number of constituent particles. Conventionally, in such cases, we prefer to use a continuous size variable, instead of the discrete particle number N, in order to represent the matter state. We are then led to a mathematical expression of the matter state represented by a direct sum of continuously many Hilbert spaces. Notice that such an expression is unitarily inequivalent to the original particle-number representation. This kind of representation space is called a "continuous-superselection-rule" space. We will see that the WFC is realized in such a continuous-superselection-rule space. More details will be explained in Chapter 4.

2.5 Future prospects

Most physicists have considered the quantum measurement problem as a very academic or a rather philosophical topic for quite a long time. Recent technological developments, however, have "pulled" the problem back to real physics. Actually, we are now able to perform, in real laboratories, those fundamental experiments which were simply considered at the *gedanken* level in the past. From this point of view, let us discuss the possible future prospects for the quantum measurement problem.

First of all, we have to pay attention to "imperfect measurements," which can be considered as an example of *mesoscopic* phenomena, and will be properly formulated and discussed in this book. Mesoscopic phenomena are nowadays rapidly expanding into new fields, requiring sophisticated technology, well beyond the fundamental interest. In such phenomena, the quantum system involved is considered to be in a partially coherent (or equivalently partially mixed) state. It is natural to expect that in the near future it will be possible to extract significant information about quantum mechanical systems through mesoscopic phenomena, i.e. imperfect measurements: We will encounter measurement processes that can only be understood by making use of the notion of imperfect measurements. In this context, one of the most important points in order to judge the validity and importance of a quantum measurement theory, both at present and in the future, is whether the theory itself enables one to discuss "imperfect measurements" or "mesoscopic phenomena." Notice that we cannot formulate a significant theory of imperfect measurements on the basis of the naive projection rule, i.e. the naive WFC, while they are properly described within the framework of the many-Hilbert-space theory, as will be explicitly shown in this book.

Among the many ways of performing quantum measurements, the so-called quantum non-demolition measurements, invented in quantum optics in order to detect photons without destroying them, will undoubtedly play a very important role in the future. We expect that this type of measurement will be extensively applied to other types of experiments. This is another future prospect for the quantum measurement problem. Quantum non-demolition measurements are not only practically useful but also theoretically very interesting, and will undoubtedly trigger the future development of experimental technique. In this context, the future measurement theory must also be examined from the point of view of future technology.

From the technical point of view, detection of weak signals will also become important in the near future. This problem will inevitably lead not only to the improvement of experimental technique, but also to the development of the theory of imperfect measurements, because we shall often encounter mesoscopic phenomena in the detection of weak signals. The quantum measurement problem is no longer of purely academic interest, but is actively linked to the development of future technology, on the basis of mesoscopic phenomena. For this reason, the future prospect of the quantum measurement problem covers possible developments of theoretical and experimental research work on imperfect measurements and mesoscopic phenomena.

Theoretically, we are also interested in the relation between dephasing and chaos, from the measurement-theoretical point of view, because the chaotic behaviour of a dynamical system is expected to provoke dephasing or irreversibility. Usually, dephasing and/or irreversibility are considered to take place in a dynamical system with a huge number of degrees of freedom. We naturally expect that a possible relationship between dephasing and chaos will shed new light on the future measurement theory, as well as on the notion of quantum chaos itself. Such an expectation, as well as its measurement-theoretical implications, should be regarded as another future perspective.

Concerning this issue, it would be interesting to examine whether a dynamical system with a few degrees of freedom can provoke dephasing leading to the wave function collapse by measurement. Of course, this idea is in contrast to the widespread opinion that only a large system can yield dephasing, but we think that it should be meaningful, in principle, because the study of chaos shows that even a classical system with a few degrees of freedom can be in chaotic motion. This problem will be discussed in Chapter 9. The idea will be realized in some cases. On the other hand, the problem of devising a chaotic detector of this kind requires us to solve the related problem of how to generate significant signals, because the possible responses that are obtainable directly from such a detector would be very weak.

As was already mentioned, any detector system should be able to perform two functions: The first is to yield signals to display results of the measurement, and the second to provoke dephasing among branch waves prepared in the spectral-decomposition step. The latter is one of the most important issues in the quantum measurement problem, to which most of this book will be devoted. As for detection, we usually expect rather weak signals from a microscopic system: they are detected by means of those amplifiers in which thermal irreversible processes take place inside a macroscopic system with a huge number of degrees of freedom. However, we would like to consider another possibility of detecting weak signals without resorting to such amplification processes. For simplicity, suppose that we are dealing with a "yes-no" experiment for the measurement of a certain dichotomic observable, as was discussed in Section *1.3*, and that we can observe and record the time-dependence of the interference term as a time-series. In this case, we can apply the so-called time-series analysis, formulated by means of Fourier analysis, to this time-variation phenomenon. The analysis may display a remarkable difference between time-series with and without signal, that is, significant information about the signal. This idea has not yet been firmly established, but is worth considering as a future prospect of the measurement problem.

The future measurement theory must prepare a sound theoretical base for new technological development, and in turn technological development will excite further theoretical studies. In spite of such a close connection to technology, however, we should never forget that the quantum measurement problem will remain for a while in the domain of foundational interest.

Chapter 2

ELEMENTS OF QUANTUM MECHANICS

We briefly review some fundamental ideas and techniques of quantum mechanics that are necessary to discuss the quantum measurement problem.

1 Wave-particle dualism

Experiments tell us that a quantum mechanical particle has both the wave and particle nature: This is called *wave-particle dualism*. The "particle" is a wave in the sense of showing interference phenomena, and is a particle by virtue of its characteristic of being an indivisible entity. The wave nature (specified by wave number k and frequency ω) and the particle nature (specified by momentum p and energy E) are related to each other by the de Broglie relations

$$k = \frac{p}{\hbar}, \quad \omega = \frac{E}{\hbar}, \tag{2.1}$$

$\hbar = 1.054 \times 10^{-34} \text{Js}$ being the Planck constant. Remember that the wave-particle dualism is an experimental fact, which is indispensable and sufficient to construct the whole theory of quantum mechanics.

1.1 Experimental facts

The wave nature is brought to light by an interference experiment, for example, of the Young type. Look again at Fig. 1.3. The incident wave ψ_0 is decomposed, by the double slit barrier, into two branch waves, ψ_a and ψ_b, which are then superposed as in ψ_I of (1.10) downstream from the barrier. Finally, we observe the particle intensity, proportional to (1.11), on a remote screen. The last term in the r.h.s. of (1.11) is called the *interference* term. The surface of the screen is covered with a large number of small detectors, by which we can observe the particle distribution as given by (1.11). (For the position measurement on the screen, refer to the discussion in Sections *1.2* in Chapter 1 and *2.2* in Chapter 4.) Needless to say, if we observe a beautiful interference pattern (as shown by P in Fig. 1.3) on the screen, we can conclude that the particle discloses its wave nature. The wave nature of quantum mechanical particles was recognized in this way. As for the wave nature, however, there is no distinction between classical and quantum mechanical waves, and the interference phenomena are explained by means of (i) the superposition law (1.10) and (ii) the intensity proportional to (1.11). The de Broglie formulae (2.1) were also verified through this kind of experiment.

The essential difference between the quantum mechanical wave-particle dualism and the classical wave nature becomes clear for a beam of very weak intensity.

When the beam intensity is gradually weakened, the interference pattern for quantum mechanical particles begins to collapse in the way shown in Fig. 2.1(a), while the pattern for classical waves is unaltered, although its amplitude is continuously reduced. Eventually, if we observe the screen for a

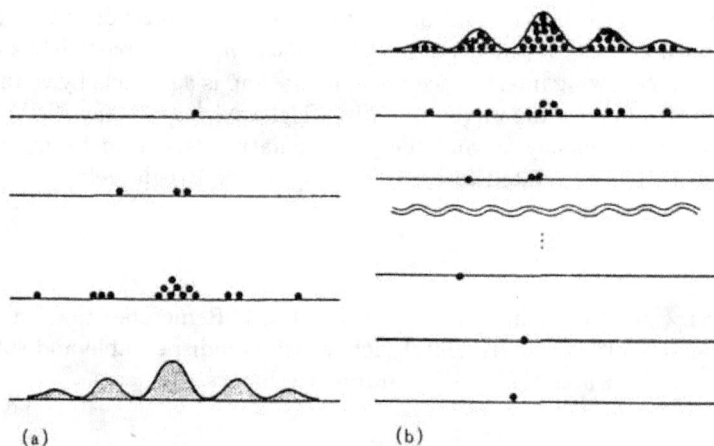

(a)   (b)

Fig. 2.1. The experimental perspective of the wave-particle dualism: (a) Observation of the particle nature by weakening the beam intensity (from the bottom to the top). (b) Observation of the wave nature, i.e. reproduction of the interference pattern by accumulating many single spots (from the bottom to the top).

short time, *only a single spot* appears on it, as in Fig. 2.1(a), for an extremely weak beam. This is the quantum-*particle* nature, in the sense that one particle is never "captured" by two or more detectors at the same time. In other words, the particle nature of quantum mechanical objects never manifests itself in more than one place at a time: particles are indivisible.

At a first glance one might think that the wave nature has disappeared in the above observation of a single spot, but the "particle" never "forgets" its wave nature, even in this case. In fact, the accumulation of many single spots made by many independent particles will reproduce the interference pattern on the screen. This is pictorially shown in Fig. 2.1(b), in which the accumulated distribution has the same form as in the interference pattern in Fig. 1.3. Note that an incident beam of strong intensity can perform this accumulation process in a very short time.

We have to emphasize that the above is *not* to be viewed as a gedanken experiment, but can be observed in actual experiments: Nowadays, technology allows us to consider concrete physical situations in which particles are recorded, *one by one*, by a macroscopic experimental setup (detector). The beautiful photograph in Fig. 2.2 clearly shows the wave-particle dualism of the electron by means of the accumulation of many single spots, corresponding to independent electrons, in an interferometry experiment performed by Tonomura. See Ref. 190. A quantum mechanical particle subject to the wave-particle dualism is described in terms of a wave function. The above experimental fact of the wave-particle dualism, with the de Broglie relations (2.1), suggests that the quantum mechanical wave function of a particle with a definite momentum p is represented by the plane wave

$$u_p(r) = \frac{1}{\sqrt{(2\pi\hbar)^3}} \exp\left[\frac{i}{\hbar} p \cdot r\right] ; \tag{2.2}$$

in other words, a particle with momentum p is described by a wave with de Broglie wavelength $\lambda = 2\pi\hbar/p$. The numerical coefficient of (2.2) is chosen so as to give

$$(u_p, u_{p'}) \equiv \int d^3r \, u_p^*(r) u_{p'}(r) = \delta^3(p - p'), \quad \int d^3p \, u_p(r) u_p^*(r') = \delta^3(r - r'). \tag{2.3}$$

The other de Broglie relation in (2.1) also suggests that the quantum mechanical state specified by a definite energy E is described by the time factor

$$\exp\left[-\frac{i}{\hbar} Et\right] . \tag{2.4}$$

The first relation, (2.2), will be accepted as one of the fundamental postulates of *wave mechanics* (i.e. *quantum mechanics*), and the second, (2.4), will be derived within the theoretical framework of wave (quantum) mechanics, as will be shown later.

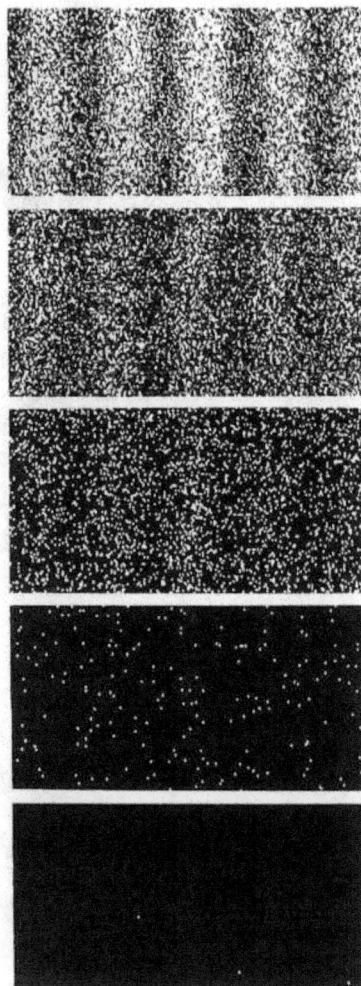

Fig. 2.2. Observing the wave-particle dualism by making use of an electron beam. We clearly observe the reproduction of the interference pattern by experimentally accumulating independent particle spots: Look at the photographs from the bottom (independent spots) to the top (interference pattern). (Courtesy of A. Tonomura.)

1.2 Probabilistic interpretation

As is well known, we can consistently understand the above experimental facts by means of a purely probabilistic interpretation of the wave function: $P = |\psi(x,t)|^2$ is proportional to the probability of finding a particle at a space-point x at time t, when it is in a state represented by ψ. Under this assumption, many particles can arrive around the peaks of P, because probabilities are high, and only a few particles can reach its valleys because probabilities are lower. This is called the "probabilistic interpretation" of the wave function.

We are thus led to *quantum mechanics* (wave mechanics), based on the postulates that the wave function ψ, representing a particle state, is subject to (i) formula (2.2), based on the de Broglie relations (2.1), (ii) the superposition principle, and (iii) the probabilistic interpretation. In addition to these, we must assume (iv) the Schrödinger equation as the dynamical postulate. Details will be given in the next section.

Suppose we find a particle at point X on the screen at time t in the interference experiment of Fig. 1.3. On the basis of the probabilistic interpretation, one can state that the particle state immediately after the observation is represented by a wave function ψ_X, distributed only around X: The measurement has caused the following sudden change of the wave function

$$\psi \rightarrow \psi_X \qquad (2.5)$$

at the instant t. This change is often called "wave function collapse" by measurement, and its general version was already given in (1.7). We have already explained that the wave function collapse by measurement is not formulated by such a transition of the projection type but rather by erasing the phase correlation among the branch waves. We called (2.5) or (1.7) the naive WFC. We shall properly formulate the wave function collapse in Chapters 3 and 4.

We should notice that, if we find the particle at X' different from X, then the naive WFC corresponds to $\psi \rightarrow \psi_{X'}$. For the *same* initial state ψ, we do not know which change will take place, whether $\psi \rightarrow \psi_X$ or $\psi \rightarrow \psi_{X'}$. This means that the wave function collapse is not a causal wave motion, leading to a continuous "shrinking" of ψ into ψ_X or $\psi_{X'}$, but is rather an *acausal* and *purely probabilistic* event. Quantum mechanics only predicts that the probability of finding each event is proportional to $|\psi(X)|^2$ or $|\psi(X')|^2$.

Furthermore, we should remark that the WFC cannot be described by the Schrödinger equation of the object particle, which produces only causal changes. This gives rise to the serious question of whether quantum mechanics is a self-contained theory, because the measuring process on the screen in Fig. 1.3, which is an acausal process, does not seem to be describable by quantum mechanics itself (i.e. the Schrödinger equation). This question engenders a primitive form of the so-called measurement problem: Can quantum mechanics describe the measuring process? We will formulate the measurement problem more rigorously later.

We have already explained, in the preceding chapter, that the above interference experiment can also be regarded as a "yes-no experiment" to determine the particle path, a or b. We shall not repeat this explanation, but remark again that the experimental process should be characterized by quantum dephasing between two branch waves, ψ_a and ψ_b. In all the cases considered, the observation or determination of the particle path results in (1.12) on the screen, which is characterized by the lack of the interference term. This means that if we observe the particle nature, we cannot see the wave nature, characterized by the interference pattern on the screen. On the other hand, we know that if we remove all detectors at slits a and b, then the interference pattern on the screen is recovered. This means that if we observe the wave nature, we cannot see the particle nature. In other words, the wave and particle natures are mutually complementary in Bohr's sense.

Against this conclusion, Einstein once claimed that we could determine the particle path without completely destroying the interference pattern. He proposed a gedanken experiment similar to the experiment considered above. As is well known, Bohr countered this objection with a qualitative discussion based on the uncertainty principle, by proving that the determination of the particle path inevitably destroys the interference pattern. Many physicists, for a long time, have been considering this issue along Bohr's line of thought. A few years ago Wootters and Zurek,[207] by applying the quantum theory of scattering, examined whether the observation of the particle path by means of a one-atom detector destroys the interference pattern. Their conclusion is that even if the particle's probabilities of passing through slits a and b are meaningfully different from each other (that is, the particle path can be determined in a probabilistic sense), the interference pattern is not completely destroyed. One might think that their conclusion supports Einstein's claim and not Bohr's. This would not be correct: They have shown that a one-atom detector with a few degrees of freedom is not sufficient to determine the particle path, but can only yield an *imperfect* measurement. We shall see later that perfect measurements can be realized by means of a macroscopic detector with a huge number of degrees of freedom. After this analysis Zurek proposed the so-called environment theory.[211] Summarizing, we agree with Bohr's qualitative arguments.

On the basis of this line of thought, we conclude that if we place a *perfect* detector at a slit, we can surely observe the particle nature but not the wave nature. If, instead, we remove the detector, we can surely see the wave nature but not the particle nature. The act of placing or removing the detector is to be ascribed to the arbitrary will of the observer. On the other hand, in this way, the phenomenon of being a "wave" or a "particle" appears as if it were subject

to the human will. This provoked epistemological arguments as to whether
a quantum phenomenon is objective or not. The proposal of delayed-choice
experiments (to be analyzed in the following subsection) sharply tackled this
point.

Another interesting (or rather paradoxical) point in the detection process
(2.5) on the screen is the fact that we have found the particle only at a space-
point X although the wave function, immediately before the detection, spreads
over a macroscopically wide region and interacts with many finely spaced detectors distributed on the screen. The "interaction" does not always yield the
"event." We have already seen a similar behaviour of the wave function in the
case of negative-result measurements,[90,168] as was discussed in Section *1.3* of
Chapter 1. However, it is not correct to think that one fraction of the wave
function interacts with a macroscopic apparatus (or one of its local systems),
independently of its other fractions, just like a classical wave. This is one of the
very confusing characteristics of quantum mechanics, namely the appearance
of the *indivisibility* of quantum systems in Bohr's sense, and is closely related
to the *nonlocal long-distance correlations* of the wave function. In this context,
it is useful to refer to the Einstein-Bohr debate.[53,28]

1.3 Delayed-choice experiment

We can observe the wave-particle dualism by another interesting phenomenon,
named "delayed-choice experiment," which was first proposed by Wheeler[203]
and experimentally observed by Alley, Walther and others[2,75] for photons and
by Kawai *et al.*[97] for neutrons. Refer to Fig. 2.3, in which we have denoted
a semi-transparent mirror with a slashed circle and a perfect mirror with a
simple slash. We send light pulses of very low intensity, one by one, to the
first semi-transparent mirror M_1. A light pulse contains less than one photon
on the average. Each light pulse is divided by M_1 into two branch waves, one
running in channel A and the other in channel B. The distance between the
first semi-transparent mirror and the perfect mirrors is much longer than the
size of the light pulse.

In the case of Fig. 2.3(a), the two branch waves (relative to one photon)
are separately forwarded to detectors D_A and D_B. Since one pulse contains
less than one photon, if D_A (D_B) detects one photon, D_B (D_A) does not. This
is due to the particle indivisibility: One photon cannot be captured by two or
more detectors. This experiment is nothing but the observation of the particle
nature.

In the case of Fig. 2.3(b), a second semi-transparent mirror M_2 is inserted,
so that the two branch waves are superposed to yield two waves running toward
D_A and D_B, respectively. A semi-transparent mirror shifts the phase of the
reflected wave by $\pi/2$ but does not modify the phase of the transmitted wave.

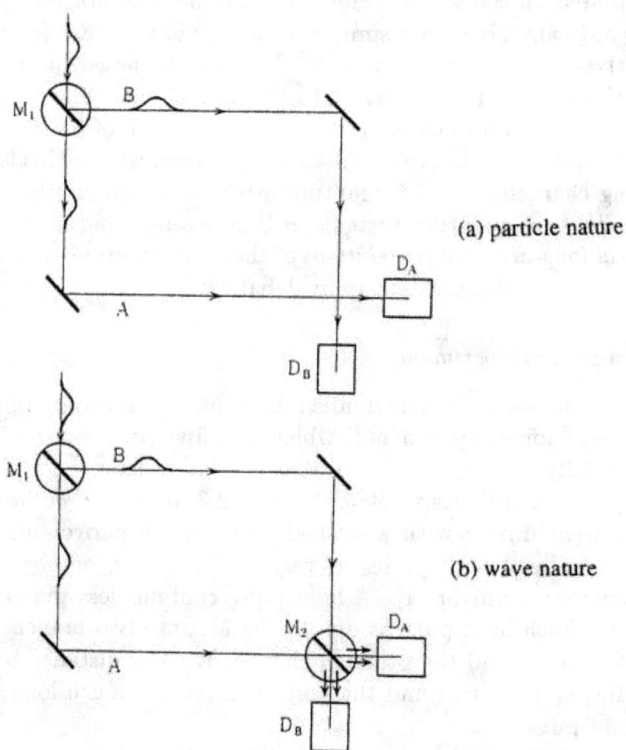

(a) particle nature

(b) wave nature

Fig. 2.3. Delayed-choice experiment.

Consequently, the wave going to D_A vanishes and does not generate a signal, because the two waves coming from channels A and B are out of phase with each other, while the wave going to D_B does not vanish and can produce a signal, because the two waves coming from channels A and B are in phase. This experiment corresponds to the observation of the wave nature. The experiment schematized by Figs. 2.3(a) and (b) is a smart verification of the wave-particle dualism. However, Wheeler's purpose was not to observe the wave-particle dualism itself, but rather to stress a very peculiar feature of quantum mechanical measurements. The observer has to make one of the following two choices *after* the light pulse has passed through the first semi-transparent mirror M_1: The first choice is "to insert" the second semi-transparent mirror M_2, while the second choice is "to do nothing." Because the choice is made only after the passage of the light pulse through the first semi-transparent mirror, this experiment is called a delayed-choice experiment. The photon behaves as a "wave" for the first choice, but behaves as a "particle" for the second choice. The photon does not know, immediately after passing through the first semi-transparent mirror, whether it will have to behave as a "wave" or a "particle" later. The very phenomenon of being a "wave" or "particle" is not definite before the detectors detect the photon, and will be determined according to the observer's will.

This experiment was performed,[2,75] yielding the result Wheeler expected. Wheeler claimed therefore that there is "no phenomenon" before "recording."

If this statement were excessively extended, after several logical steps, one might reach the conclusion that the universe itself was waiting for the appearance of the human being in order to be "recognized," because the universe as a "phenomenon" should not exist before the human being observed and recorded it. This aspect would be an extreme form of the so-called *anthropic principle*. The reader should recall some rather old philosophical arguments about the uncertainty principle, in which some physicists and philosophers thought that human intention could govern the quantum world. Such a trend has often appeared in discussions on the measurement problem (see, for example, the von Neumann-Wigner approach in Section 2.1 of Chapter 1 and Section 2 of Chapter 3).

Wheeler's statement can be justified only if we rigidly restrict the notion of "phenomenon" to "wave" and "particle." On the other hand, however, we can also think of a quantum mechanical particle as a physical entity endowed with the wave and particle nature as its two fundamental aspects. In this context, Wheeler's choice of inserting the second semi-transparent mirror or not is merely a technique that enables us to pick up one of these two aspects. Wheeler's statement logically counters the standpoint (engendered by a classical view) that "wave" and "particle" are independently (and exclusively) connected to "physical entities." Nevertheless, one may say that the concept of "quantum mechanical entity" is not yet as clear as the classical one. In this book we shall not enter into this kind of epistemological argument.

2   Formulation of quantum mechanics

Quantum mechanics is formulated in terms of "states" (or wave functions) and "observables," together with their physical interpretation, and a "dynamical principle."

2.1   Fundamental postulates

Wave mechanics (i.e. the primitive form of quantum mechanics) starts from the notion of *state*.

State: A state of motion of a dynamical system with n degrees of freedom at time t is described in terms of the *wave function* $\psi(x, t)$, which depends on $x = (x_1, x_2, \cdots, x_n)$, x representing a point in configuration space: For example, $n = 3$ for one-particle, $n = 6$ for two-particle, ... and $n = 3N$ for N-particle systems, respectively, if we neglect any kind of constraints. Notice that we deal with x as if it belonged to a one-dimensional space. For simplicity, let us call such a state "ψ-state." Furthermore, if the particles have other quantum numbers, the wave function will depend on them, as well. For a while, however, we shall suppress the dependence on these additional quantum numbers. We usually call $\psi(x, t)$ (depending on x) the "*position* representation" of a state.

Fundamental postulates in wave mechanics: The fundamental postulates of wave mechanics on the wave function $\psi(x, t)$ are the following.

1. **Definite momentum state:** It is specified by the momentum $p = (p_1, p_2, \cdots, p_n)$ and is written as

$$u_p(x) = \frac{1}{(\sqrt{2\pi\hbar})^n} e^{ip \cdot x/\hbar}, \tag{2.6}$$

   in the position representation, where $p \cdot x = \sum_{i=1}^{n} p_i x_i$. We shall see later that this is the transformation function (or kernel) between the position and momentum representations.

2. **Superposition principle:** If the states of motion described by the wave functions, ψ_1, ψ_2, $\cdots$, are simultaneously realized, the quantum state is represented by their sum

$$\psi = \psi_1 + \psi_2 + \cdots. \tag{2.7}$$

3. **Probabilistic interpretation**: The probability of finding the particle in a small domain (of volume $dx \equiv \prod_{i=1}^{n} dx_i$) around x at time t is proportional to

$$|\psi(x,t)|^2 dx. \tag{2.8}$$

In this context, we shall call $\psi(x,t)$ the probability amplitude. Obviously, $|\psi(x,t)|^2$ itself becomes the probability density of finding the particle at x and t, provided that the wave function is normalized as

$$\int dx |\psi(x,t)|^2 = 1. \tag{2.9}$$

The probabilistic interpretation holds on the basis of the probability ensemble that consists of many experimental results on identical quantum systems, each of which is represented by the same wave function. In the case of scattering processes, many incident particles sent by the incident beam will constitute the probability ensemble.

4. **Dynamical principle**: The wave function obeys the Schrödinger equation

$$i\hbar \frac{\partial}{\partial t}\psi(x,t) = H^x \psi(x,t), \tag{2.10}$$

where H^x is a linear self-adjoint operator obtained from the classical Hamiltonian by the rule given below.

The first two postulates enable us to represent a general quantum mechanical state, $\psi(x,t)$, as a superposition of definite momentum states:

$$\psi(x,t) = \int dp\, u_p(x)\tilde{\psi}(p,t), \qquad \left(dp \equiv \prod_{i=1}^{n} dp_i\right), \tag{2.11}$$

where $\tilde{\psi}(p,t)$ is called the wave function in the *momentum* representation. Mathematically, we know that (2.11) is a Fourier transformation and we have its inverse transformation

$$\tilde{\psi}(p,t) = \int dx\, u_p^*(x)\psi(x,t). \tag{2.12}$$

Both $\psi(x,t)$ and $\tilde{\psi}(p,t)$ give mathematically and physically equivalent descriptions of the quantum mechanical state.

Abstract representation and quantum mechanics: By virtue of the superposition principle, we know that the wave function is a member of a vector space, for example, a Hilbert space. Remember that an L^2 space (one possible

Hilbert space) includes the definition of inner product, which is convenient to formulate the postulate of the probabilistic interpretation.

From this point view, in order to describe a quantum state, it is convenient to introduce an abstract vector space, in which a dynamical state is represented by an abstract vector $|\psi_t\rangle$ instead of a c-number function $\psi(x,t)$. We usually call $|\psi_t\rangle$ "ket" vector and its adjoint $\langle\psi_t|$ "bra" vector, according to Dirac's nomenclature. Their interrelationship is

$$|\psi_t\rangle = \int dx|x\rangle\psi(x,t), \quad \langle\psi_t| = \int dx\psi^*(x,t)\langle x|, \qquad (2.13)$$

where $|x\rangle$ stands for an eigenvector of the abstract *position* operator $\hat{x}$, belonging to the eigenvalue x, and $\langle x|$ for its adjoint. They obey the following eigenvalue equations and orthonormal completeness conditions:

$$\hat{x}|x\rangle = x|x\rangle, \quad \langle x|\hat{x} = \langle x|x, \qquad (2.14)$$

$$\langle x|x'\rangle = \delta(x-x'), \quad \int dx|x\rangle\langle x| = \hat{1}, \qquad (2.15)$$

where $\delta(x) \equiv \prod_{i=1}^{n}\delta(x_i)$. Because of (2.15), we obtain

$$\psi(x,t) = \langle x|\psi_t\rangle, \quad |\psi_t\rangle = \int dx|x\rangle\langle x|\psi_t\rangle, \qquad (2.16)$$

$$\langle\phi|\psi\rangle = \int dx\phi^*(x,t)\psi(x,t) = (\phi,\psi). \qquad (2.17)$$

Observe also that the definite momentum state $u_p(x)$ in (2.6) is expressed as

$$u_p(x) = \langle x|p\rangle = \langle p|x\rangle^* \qquad (2.18)$$

in the abstract representation and therefore the wave function $\tilde{\psi}(p,t)$ in (2.11) or (2.12) is nothing but the momentum representation of the abstract vector

$$\tilde{\psi}(p,t) = \int dx\langle p|x\rangle\psi(x,t) = \int dx\langle p|x\rangle\langle x|\psi_t\rangle = \langle p|\psi_t\rangle. \qquad (2.19)$$

By making use of the abstract representation, we can smoothly transfer wave mechanics to a general formulation of quantum mechanics.

Needless to say, a quantum mechanical state is represented by an abstract vector $|\psi_t\rangle$ in the abstract vector space. In the abstract representation, the above probabilistic postulate must be generalized as follows:

$3'$. **General probabilistic interpretation:** The probability of finding the ϕ-state in the ψ-state is proportional to

$$|\langle\phi|\psi\rangle|^2 \qquad (2.20)$$

In this context, we call $\langle\phi|\psi\rangle$ itself the "probability amplitude" of finding the ϕ-state in the ψ-state. In order to incorporate the probability amplitude, we have to require that the abstract vector space, as the representation space of quantum mechanics, is endowed with the inner product $\langle\phi|\psi\rangle$ (a complex number), between the two member vectors $|\psi\rangle$ and $|\phi\rangle$. We can generalize wave mechanics to ordinary quantum mechanics, by making use of such an abstract vector representation.

Observables: A definite momentum state (or the transformation kernel) (2.6) is an eigenfunction of the operator $-i\hbar(\partial/\partial x_i)$ belonging to the eigenvalue p_i. In other words, the components of coordinates and momenta, as dynamical quantities in quantum mechanics, are represented by the abstract operators $\hat{x}_i$ and $\hat{p}_i$ ($i = 1, 2, \cdots, n$) on the abstract vector space and are subject to the commutation relations

$$[\hat{x}_i, \hat{p}_j] = i\hbar\delta_{ij}, \quad [\hat{x}_i, \hat{x}_j] = [\hat{p}_i, \hat{p}_j] = 0. \tag{2.21}$$

We can easily understand that if we take the *position* representation in which $\hat{x}_i$ is represented by a c-number variable x_i, the commutation relation (2.21) immediately enables us to write the ith component of the momentum operator as $-i\hbar(\partial/\partial x_i)$ in this position representation. This operator has (2.6) as its eigenfunction belonging to eigenvalue p_i. In other words, the essential properties of the dynamical quantities are summarized by the commutation relations (2.21). For this reason, we often call (2.21) the *quantum condition*.

In general, a dynamical quantity in quantum mechanics, called an *observable*, is assumed to be represented by a *linear self-adjoint* operator, say $\hat{F} = F(\hat{x}, \hat{p})$, transforming an abstract vector, say $|\psi\rangle$, into another one, say $|\phi\rangle$, in the above vector space

$$|\phi\rangle = \hat{F}|\psi\rangle. \tag{2.22}$$

The operator has its eigenstate $|\nu\rangle$ belonging to the eigenvalues λ_ν

$$\hat{F}|\nu\rangle = \lambda_\nu|\nu\rangle. \tag{2.23}$$

Mathematically, we know that the eigenvalues are real and the eigenstates make up an orthonormal set.

Postulates on observables:

1. **Linearity.** We can understand the linearity on the basis of the superposition principle.

2. **Measured values.** Physically, we postulate that we obtain *one* of the eigenvalues when we perform a measurement of the dynamical quantity

represented by $\hat{F}$. As for the measurement, the orthogonality implies that when we obtain one value, say λ_ν, we never obtain another value, say $\lambda_{\nu'}$ ($\nu' \neq \nu$). We can easily understand such a situation, based on the orthogonality $\langle \nu' | \nu \rangle = 0$ and the generalized probabilistic postulate (2.20).

3. **Completeness.** Furthermore, we have to assume that $\{|\nu\rangle\}$ must be complete in the sense that an arbitrary state $|\psi\rangle$ can be expanded in a series of eigenstates as follows:

$$|\psi\rangle = \sum_\nu c_\nu |\nu\rangle, \quad c_\nu \equiv \langle \nu | \psi \rangle. \tag{2.24}$$

Physically, the completeness condition means that every measurement of $\hat{F}$, performed on a dynamical system in a state $|\psi\rangle$, necessarily yields one λ as a measured value. Obviously, c_ν is the probability amplitude of finding the νth eigenstate in the ψ-state, which is subject to

$$\langle \psi | \psi \rangle = \sum_\nu |c_\nu|^2 = 1, \tag{2.25}$$

provided that $|\psi\rangle$ is normalized as in (2.9).

Other details of the operator nature are given through the commutation relations, together with the additional prescription that the ordering of $\hat{x}$ and $\hat{p}$ in $\hat{F}$ must be arranged so as to yield a self-adjoint operator such as, for example, by means of a symmetrization procedure.

We define the matrix element of $\hat{F}$ with respect to a complete vector set $\{|i\rangle\}$ by

$$F_{ij} \equiv \langle i | \hat{F} | j \rangle. \tag{2.26}$$

The matrix F_{ij} has the same property as the original abstract operator $\hat{F}$, and is called its matrix representation. If we take the set of the eigenvectors (2.23) for $\hat{F}$, we get the following diagonal matrix

$$\langle \nu | \hat{F} | \nu' \rangle = \lambda_\nu \delta_{\nu\nu'}. \tag{2.27}$$

This is nothing but the diagonal representation of $\hat{F}$, which is obtained on the vector set $\{|\nu\rangle\}$. We call this the F-diagonal representation. In the case of continuous eigenvalues, Kronecker's delta must be replaced with Dirac's delta function. For another observable, say $\hat{G}$, having eigenvalues γ_μ and eigenvectors $|\mu\rangle$, we can take another diagonal representation to give

$$\langle \mu | \hat{G} | \mu' \rangle = \gamma_\mu \delta_{\mu\mu'}. \tag{2.28}$$

Both representations are related to each other through the following transformation formulae

$$|\nu\rangle = \sum_\mu |\mu\rangle\langle\mu|\nu\rangle, \quad |\mu\rangle = \sum_\nu |\nu\rangle\langle\nu|\mu\rangle \quad (2.29)$$

and their adjoints, where we call $\langle\mu|\nu\rangle$ and its adjoint the *transformation function*. The *position* representation is nothing but the x-diagonal one. The matrix elements of the momentum operators and of $\hat{F} = F(\hat{x}, \hat{p})$ in the x-diagonal representation are easily derived from (2.21) in the following way:

$$\langle x|\hat{x}_i|x'\rangle = x_i\delta(x - x'), \quad (2.30)$$

$$\langle x|\hat{p}_i|x'\rangle = -i\hbar\frac{\partial}{\partial x_i}\delta(x - x'), \quad (2.31)$$

$$\langle x|\hat{F}|x'\rangle = F\left(x, -i\hbar\frac{\partial}{\partial x}\right)\delta(x - x'). \quad (2.32)$$

We can also introduce the *momentum* representation, i.e. the p-diagonal one, by making use of the vector set $\{|p\rangle\}$ which is subject to the orthonormal completeness condition obtained by replacing x with p in (2.15). The matrix elements of the position operators and of $\hat{F} = F(\hat{x}, \hat{p})$ in the p-diagonal representation are derived from (2.21) in the following way:

$$\langle p|\hat{p}_i|p'\rangle = p_i\delta(p - p'), \quad (2.33)$$

$$\langle p|\hat{x}_i|p'\rangle = i\hbar\frac{\partial}{\partial p_i}\delta(p - p'), \quad (2.34)$$

$$\langle p|\hat{F}|p'\rangle = F\left(i\hbar\frac{\partial}{\partial p}, p\right)\delta(p - p'). \quad (2.35)$$

The transformation function (or kernel) between the x- and p-representations is given by (2.6) or (2.18).

The introduction of the above abstract operators, together with the commutation relations (2.21), completes the general formulation of quantum mechanics. Wave mechanics is merely the *position* representation of quantum mechanics.

The Hamiltonian (energy) operator $\hat{H}$ is also introduced from the classical Hamiltonian $H(x, p)$ as

$$\langle x|\hat{H}|x'\rangle = H^x\delta(x - x'), \quad H^x = H\left(x, -i\hbar\frac{\partial}{\partial x}\right). \quad (2.36)$$

The dynamical principle of quantum mechanics is given by the abstract form of the Schrödinger equation

$$i\hbar\frac{d}{dt}|\psi_t\rangle = \hat{H}|\psi_t\rangle, \quad (2.37)$$

whose *position* representation is

$$i\hbar\frac{\partial\psi(x,t)}{\partial t} = H^x\psi(x,t), \tag{2.38}$$

where H^x is given by the second formula in (2.36).

For a time-independent Hamiltonian, (2.38) allows us to factorize the time dependence of $\psi(x,t)$ as

$$\psi(x,t) = u_E(x)e^{-iEt/\hbar}, \tag{2.39}$$

where $u_E(x)$ is a solution of the eigenvalue equation

$$H^x u_E(x) = Eu_E(x), \tag{2.40}$$

whose abstract form is

$$\hat{H}|u_E\rangle = E|u_E\rangle, \quad u_E(x) \equiv \langle x|u_E\rangle. \tag{2.41}$$

Since the eigenvalue E is real for the self-adjoint H^x, the state described by (2.39) is *stationary* in the sense that the probability $|\psi(x,t)|^2 = |u_E(x)|^2$ is independent of time. We can mathematically prove that a stationary state inevitably satisfies the stationary Schrödinger equation (2.40). Note that $u_E(x)$ is nothing but the transformation function between the H-diagonal and x-diagonal representations.

Thus we have derived the harmonic oscillation, $\exp[-iEt/\hbar]$, for an energy-definite state, as suggested by the de Broglie relations. See the remarks at the end of Section *1.1*.

2.2   *Temporal evolution and S-matrix*

We have already given the dynamical principle in the form of the abstract Schrödinger equation (2.37), in which the state vector $|\psi_t\rangle$, or the wave function $\psi(x,t)$, is exclusively responsible for the temporal evolution, while all observables $\hat{F}$ do not explicitly depend on time. Such a description of the temporal evolution is usually called the *Schrödinger picture*: Here, only the state vector is responsible for the temporal evolution, while all operators, describing observables, are time independent. Write the formal solution of (2.37) as

$$|\psi_t\rangle = \hat{U}(t,t')|\psi_{t'}\rangle, \tag{2.42}$$

where $\hat{U}(t,t')$ is the temporal evolution operator, obeying the operator Schrödinger equation and the initial condition

$$i\hbar\frac{d}{dt}\hat{U}(t,t') = \hat{H}\hat{U}(t,t'), \quad -i\hbar\frac{d}{dt'}\hat{U}(t,t') = \hat{U}(t,t')\hat{H}, \quad \hat{U}(t,t) = \hat{1}. \tag{2.43}$$

Hereafter we shall simply denote the identity operator as $\hat{1} = 1$. If the Hamiltonian operator does not explicitly depend on time t, we are led to

$$\hat{U}(t, t') = e^{-i\hat{H}(t-t')/\hbar}. \tag{2.44}$$

We shall sometimes write

$$|\psi_t\rangle = \hat{U}(t)|\psi_0\rangle, \quad \hat{U}(t) \equiv \hat{U}(t, 0), \tag{2.45}$$

where the initial condition $|\psi_0\rangle$ is set at $t = 0$.

Let us rewrite the expectation value of an observable $\hat{F}$ at time t in terms of (2.45) as

$$\langle F \rangle_t \equiv \langle \psi_t|\hat{F}|\psi_t\rangle = \langle \psi_0|\hat{U}^\dagger(t)\hat{F}\hat{U}(t)|\psi_0\rangle = \langle \psi_0|\hat{F}_t|\psi_0\rangle, \tag{2.46}$$

where

$$\hat{F}_t \equiv \hat{U}^\dagger(t)\hat{F}\hat{U}(t). \tag{2.47}$$

The description by means of a time-independent state vector $|\psi_0\rangle$ and time-dependent observables $\hat{F}_t$ is called the *Heisenberg picture*: Here the state vector is kept in the initial state $|\psi_0\rangle$, while all observables change according to the *Heisenberg* equation of motion

$$i\hbar\frac{d}{dt}\hat{F}_t = [\hat{F}_t, \hat{H}], \tag{2.48}$$

which is easily derived from (2.47) and (2.43). We can calculate the r.h.s. of this equation by making use of the canonical commutation relations

$$[\hat{x}_{i,t}, \hat{p}_{i,t}] = i\hbar\delta_{ij}, \quad [\hat{x}_{i,t}, \hat{x}_{j,t}] = [\hat{p}_{i,t}, \hat{p}_{j,t}] = 0, \tag{2.49}$$

which are easily derived from (2.21) by (2.47). As is well known, the operator equation of motion is very close to the classical one, except for the ordering of operators.

Finally, we introduce the *interaction picture* of temporal evolutions, assuming that the Hamiltonian operator can be decomposed as

$$\hat{H} = \hat{H}_0 + \hat{H}_I, \tag{2.50}$$

where $\hat{H}_0$ and $\hat{H}_I$ stand for the free and the interaction Hamiltonians, respectively. Usually, by assuming that the interaction part $\hat{H}_I$ yields small effects in comparison with the free part $\hat{H}_0$, we expand $\hat{U}(t, t')$ with respect to $\hat{H}_I$. The state vector and the temporal evolution operator in the interaction picture are defined by removing the contribution of the free evolution from both states and operators

$$|\psi_t\rangle_I = e^{i\hat{H}_0 t/\hbar}|\psi_t\rangle = \hat{U}_I(t, t')|\psi_{t'}\rangle_I, \tag{2.51}$$

$$\hat{U}_I(t, t') = e^{i\hat{H}_0 t/\hbar}\hat{U}(t, t')e^{-i\hat{H}_0 t'/\hbar}. \tag{2.52}$$

The initial condition is still kept unaltered: $\hat{U}_I(t', t') = 1$. The Schrödinger equation for the state vector and the temporal evolution operator in the interaction picture are easily found to be

$$i\hbar \frac{d}{dt}|\psi_t\rangle_I = \hat{H}_{I,t}|\psi_t\rangle_I, \tag{2.53}$$

$$i\hbar \frac{d}{dt}\hat{U}_I(t, t') = \hat{H}_{I,t}\hat{U}_I(t, t'), \quad -i\hbar \frac{d}{dt'}\hat{U}_I(t, t') = \hat{U}_I(t, t')\hat{H}_{I,t'}, \tag{2.54}$$

where $\hat{H}_{I,t} \equiv e^{i\hat{H}_0 t/\hbar}\hat{H}_I e^{-i\hat{H}_0/\hbar}$ is the interaction Hamiltonian operator in the interaction picture. We have defined the time-dependent operator in the interaction picture

$$\hat{F}_{I,t} \equiv e^{i\hat{H}_0 t/\hbar}\hat{F}e^{-i\hat{H}_0/\hbar} = \hat{U}_I(t)\hat{F}_t\hat{U}_I^\dagger(t), \qquad \hat{U}_I(t) \equiv \hat{U}_I(t, 0). \tag{2.55}$$

Summarizing, we have a time-dependent state vector $|\psi_t\rangle$ and time-independent observables $\hat{F}$ in the *Schrödinger picture*, a time-independent state vector $|\psi_0\rangle$ and time-dependent observables $\hat{F}_t$ in the *Heisenberg picture*, and a time-dependent state vector $|\psi_t\rangle_I$, driven only by the interaction Hamiltonian, and time-dependent observables $\hat{F}_{I,t}$, driven only by the free Hamiltonian, in the *interaction picture*. Here we have chosen all states so that they coincide at the initial time $t = 0$.

In the *interaction picture*, we can obtain $\hat{U}_I(t, t')$ by solving (2.54) and then we can get the *S-matrix* for the physical process concerned, by the formula

$$\hat{S} = \lim_{\substack{t \to \infty \\ t' \to -\infty}} \hat{U}_I(t, t'), \tag{2.56}$$

or equivalently,

$$e^{-i\hat{H}t/\hbar} \xrightarrow{t \to \infty} e^{-i\hat{H}_0 t/\hbar}\hat{S}. \tag{2.57}$$

As is well known, (2.54) can be equivalently replaced with the integral equation

$$\hat{U}_I(t, t') = 1 + \frac{1}{i\hbar}\int_{t'}^t dt'' \hat{H}_{I,t''}\hat{U}_I(t'', t'), \tag{2.58}$$

incorporating the initial condition. Solving this integral equation by iteration, we can obtain the perturbation series

$$\hat{U}_I(t, t') = \sum_{n=0}^\infty \frac{1}{(i\hbar)^n}\int_{t'}^t dt_1 \hat{H}_{I,t_1}\int_{t'}^{t_1} dt_2 \hat{H}_{I,t_2}\cdots\int_{t'}^{t_{n-1}} dt_n \hat{H}_{I,t_n}. \tag{2.59}$$

3 Density matrix

Roughly speaking, a quantum mechanical state described by a single wave function is called a *pure state*, while a state represented by means of a statistical ensemble composed of many pure states is called a *mixed state*. We shall redefine both notions of state by making use of the density matrix. The density-matrix method is indispensable to express mixed states.

First of all, let us consider the dyadic product $|a\rangle\langle b|$ of a ket vector $|a\rangle$ and a bra vector $\langle b|$. This is an operator yielding the ket vector $|a\rangle\langle b|\psi\rangle$ when applied to $|\psi\rangle$ from the left, and the bra vector $\langle\phi|a\rangle\langle b|$ when applied to $\langle\phi|$ from the right. Note that $\langle b|\psi\rangle$ and $\langle\phi|a\rangle$ are c-numbers. Now, in order to represent a quantum mechanical state corresponding to $|\psi\rangle$, we introduce the *density matrix*, defined by the dyadic product

$$\hat{\rho} \equiv |\psi\rangle\langle\psi|. \tag{2.60}$$

This operator is nothing but the projection operator onto the state $|\psi\rangle$ or $\langle\psi|$, because we obtain

$$\hat{\rho}|\phi\rangle = |\psi\rangle\langle\psi|\phi\rangle, \quad \langle\phi|\hat{\rho} = \langle\phi|\psi\rangle\langle\psi|, \tag{2.61}$$

by applying $\hat{\rho}$ to an arbitrary ket vector $|\phi\rangle$ from the left or to an arbitrary bra vector $\langle\phi|$ from the right, respectively. Furthermore, we can prove the projection property

$$\hat{\rho}^2 = \hat{\rho}, \tag{2.62}$$

provided that $|\psi\rangle$ is normalized

$$\langle\psi|\psi\rangle = 1. \tag{2.63}$$

For this reason, we can identify a *pure state* with a state represented by a projection-operator density matrix. The projection-operator density matrix is simply called the pure-state density matrix.

If we use a basic set of vectors $\{|u_i\rangle\}$, we can expand the state vector and the density matrix for a pure state as follows:

$$|\psi\rangle = \sum_i c_i|u_i\rangle, \quad c_i \equiv \langle u_i|\psi\rangle, \tag{2.64}$$

$$\hat{\rho} = \sum_i \sum_j \rho_{ij}|u_i\rangle\langle u_j|, \quad \rho_{ij} \equiv \langle u_i|\hat{\rho}|u_j\rangle = c_i c_j^*. \tag{2.65}$$

For the position representation, we have

$$\rho_{xx'} = \psi(x)\psi^*(x'). \tag{2.66}$$

The expectation value of F in the ψ-state is given by

$$\langle F \rangle = \langle \psi | \hat{F} | \psi \rangle = \mathrm{Tr}(\hat{F}\hat{\rho}). \tag{2.67}$$

Observe that the phase correlations among the $|u_i\rangle$s, for the pure state described by (2.64) and/or (2.65), are completely preserved, and the expectation value has the simple form (2.67).

Let us now consider a statistical ensemble composed of quantum systems, each of which is described by a state vector $|\psi_m\rangle$ with a statistical weight w_m ($\sum_m w_m = 1$). We shall call a state described by making use of this statistical ensemble a *mixed state*. For such a mixed state, the expectation value of the observable $\hat{F}$ is clearly given by

$$\langle F \rangle = \sum_m w_m \langle \psi_m | \hat{F} | \psi_m \rangle = \mathrm{Tr}(\hat{F}\hat{\rho}), \tag{2.68}$$

where

$$\hat{\rho} \equiv \sum_m w_m \hat{\rho}_m, \quad \hat{\rho}_m = |\psi_m\rangle\langle\psi_m|. \tag{2.69}$$

We have just introduced the density matrix $\hat{\rho}$ for a mixed state. In other words, a state which is represented by such a density matrix (with $w_m \neq 1$ for every m) is nothing but a *mixed state*. The mixed-state density matrix does not enjoy the projection property, because

$$\hat{\rho}^2 \neq \hat{\rho}, \tag{2.70}$$

as can be easily seen.

For the special choice $|\psi_k\rangle = |u_k\rangle$ and $w_k = |c_k|^2$, (2.69) becomes

$$\hat{\rho} = \sum_k |c_k|^2 |u_k\rangle\langle u_k|. \tag{2.71}$$

This density matrix expresses a mixed state with no phase correlations among the u_is. We can now understand the difference between (2.65) and (2.71), by just looking at them: The former keeps all the phase correlations among the branch waves, while the latter does not include any phase correlations. We can also easily recognize the difference between formulae (2.67) and (2.68) for the expectation values in *pure* and *mixed* states. The mixed state will play an important role in measurement theory, because it is necessary to describe a quantum state that has undergone quantum dephasing.

A state whose density matrix has no off-diagonal element like (2.71) shall be called a "completely mixed state." Contrary to this, an "incompletely mixed state" will be represented by a density matrix whose off-diagonal elements do not vanish. Note that the latter is also a mixed state.

Both the time-dependent pure- and mixed-state density matrices obey the following equation

$$i\hbar \frac{d}{dt}\hat{\rho}_t = [\hat{H}, \hat{\rho}_t], \tag{2.72}$$

which is easily derived from the Schrödinger equation for time-dependent state vectors. Note that we have used the same notation, $\hat{\rho}_t$, for both the time-dependent pure and mixed density matrices.

4 Grand quantum mechanics

In this chapter we have formulated quantum mechanics starting from the notion of wave function. This is a convenient introduction to elementary quantum mechanics, but is too narrow to deal with more general and complicated quantum systems. In order to allow more flexibility for the description of quantum systems, it is better to follow an axiomatic construction of quantum mechanics. In this section, we only outline the general ideas.

This approach usually starts by defining *observables* as a complete set of linear self-adjoint operators (more precisely, C^*-algebras or von Neumann rings), say

$$\hat{A}^{(1)}, \hat{A}^{(2)}, \cdots, \hat{A}^{(n)}. \tag{2.73}$$

For these observables, we introduce n real c-numbers

$$E(\hat{A}^{(1)}), E(\hat{A}^{(2)}), \cdots, E(\hat{A}^{(n)}), \tag{2.74}$$

and a definite mechanism to relate these c-numbers to the expectation values of the observables obtained by measurement. This mechanism entails the notion of a *state* describing the quantum mechanical system.

Such a state is mathematically represented by a "large" vector Ψ, belonging to a "large" Hilbert space which is a direct sum of "small" Hilbert spaces

$$\mathcal{H} = \mathcal{H}_1 \oplus \mathcal{H}_2 \oplus \cdots \mathcal{H}_N \oplus \int \oplus d\mu(\zeta)\mathcal{H}(\zeta), \tag{2.75}$$

where $\mu(\zeta)$ is a measure on a continuous parameter ζ to be determined, which depends on the quantum system concerned. In this way the above-mentioned expectation values are given by

$$E(\hat{A}^{(k)}) \equiv (\Psi, \hat{A}^{(k)}\Psi), \quad (k = 1, 2, \cdots, n). \tag{2.76}$$

Here Ψ is decomposed as

$$\Psi = \psi_1 \oplus \psi_2 \oplus \cdots \psi_N \oplus \int \oplus d\mu(\zeta)\psi(\zeta), \tag{2.77}$$

where ψ_k belongs to the small Hilbert space $\mathcal{H}_k$, and so on. The above expectation value is decomposed as

$$(\Psi, \hat{A}^{(k)}\Psi) = (\psi_1, \hat{A}_1^{(k)}\psi_1) + (\psi_2, \hat{A}_2^{(k)}\psi_2) + \cdots + (\psi_N, \hat{A}_N^{(k)}\psi_N)$$
$$+ \int d\mu(\zeta)(\psi(\zeta), \hat{A}^{(k)}(\zeta)\psi(\zeta)). \qquad (2.78)$$

In this mathematical procedure the inner product of two "large vectors," Ψ and Φ, is given by

$$(\Phi, \Psi) = (\phi_1, \psi_1) + (\phi_2, \psi_2) + \cdots + (\phi_N, \psi_N) + \int d\mu(\zeta)(\phi(\zeta), \psi(\zeta)). \quad (2.79)$$

Each small observable $\hat{A}_m^{(k)}$ $(k = 1, 2, \cdots, n)$, is defined on the corresponding small Hilbert space $\mathcal{H}_m$, in the sense that when it is applied to a small vector in $\mathcal{H}_m$, it yields another small vector in the same Hilbert space; the "large observable" $\hat{A}^{(k)}$ $(k = 1, 2, \cdots, n)$, when applied to a large vector yields another large vector, expressed as a direct sum, such as (2.77). This is accomplished by defining a "large observable" $\hat{A}^{(k)}$ $(k = 1, 2, \cdots, n)$

$$\hat{A}^{(k)} = \hat{A}_1^{(k)} \oplus \hat{A}_2^{(k)} \oplus \hat{A}_N^{(k)} \oplus \int \oplus d\mu(\zeta)\hat{A}^{(k)}(\zeta). \qquad (2.80)$$

A "large density matrix" Ξ is also defined by the direct sum of "small density matrices,"

$$\hat{\Xi} = \hat{\rho}_1 \oplus \hat{\rho}_2 \oplus \hat{\rho}_N \oplus \int \oplus d\mu(\zeta)\hat{\rho}(\zeta), \qquad (2.81)$$

under the trace-class assumption

$$\text{Tr}\hat{\Xi} = \sum_{m=1}^{N} \text{Tr}\rho_m + \int d\mu(\zeta)\text{Tr}\rho(\zeta) < +\infty. \qquad (2.82)$$

This mathematical procedure is called the GNS (Gelfand-Naimark-Segal) construction method.[a]

The above notion of *state* covers a wide class of states including mixed states. For example, if we take

$$\psi_m = \sqrt{w_m}u_m \qquad (2.83)$$

in (2.77), we obtain

$$E(\hat{A}^{(k)}) = \sum_{m=1}^{N} w_m \langle u_m|\hat{A}^k|u_m\rangle + \int d\mu(\zeta)w(\zeta)\langle u(\zeta)|\hat{A}(\zeta)|u(\zeta)\rangle$$
$$= \text{Tr}(\hat{A}^{(k)}\hat{\Xi}). \qquad (2.84)$$

[a]There are many papers and books on the GNS construction. See, for example, Araki's textbook and papers, Ref. 6.

We easily understand, from the point of view expressed by (2.69), that the density matrix $\hat{\bar{\Xi}}$ unequivocally describes a mixed state.

Conventionally, however, it is natural to identify the index k and the variable ζ, respectively, with the number of constituents and a continuous size parameter of a given small system. As we have seen in the preceding sections, if many local systems of the detector, characterized by different numbers of constituent particles and/or continuously different sizes, are involved in one experimental run of a measurement, we have to represent the whole state of the detector in terms of a "large density matrix" or a "large state vector" of the kind described above. Later (in Section 2.1 of Chapter 4), we shall introduce such a continuous parameter as the size variable of the local systems of the detector, for example, linear dimensions, volume, current intensity, magnetic flux and so on. The continuous part of the direct sum will become important in the measurement theory, as will be shown later, and will be called the *continuous-superselection-rule* space, while the discrete part of the direct sum will be called the *discrete-superselection-rule* space. We know that a continuous-superselection-rule space is suitable for the representation of quantum systems with infinite degrees of freedom, such as those that are usually dealt with within the framework of quantum field theory or condensed matter physics. For this reason we think that the macroscopic instruments appearing in measurement processes should be represented in such a space.

Mathematically, we also know that the continuous-superselection-rule space is endowed with a so-called *center*, containing those observables which commute with all other observables. Such systems must be governed by classical dynamics. In this context, the above scheme, making use of a "large Hilbert space" theory, can account for both classical and quantum systems, so that we may call the present scheme of dynamics, based on the "large Hilbert space," *grand quantum mechanics*. For these reasons, we think that the measurement theory should be studied within the theoretical framework of grand quantum mechanics.

Another important mathematical feature is the limiting procedure $N \rightarrow \infty$, according to which a Hilbert space $\mathcal{H}_N$ will belong to an appropriate continuous-superselection-rule space if an appropriate topology in the "large space" is chosen. Symbolically, this procedure is expressed by

$$\lim_{N \to \infty} \mathcal{H}_N \subset \int \oplus d\mu(\zeta)\mathcal{H}(\zeta). \tag{2.85}$$

In this context, remember that the infinite N limit naturally leads to the mathematical representation of the continuous-superselection-rule space.

For example, let us consider the average spin operators defined by

$$\hat{\Sigma}_k \equiv \frac{1}{N} \sum_{n=1}^{N} \hat{\sigma}_k^{(n)}, \qquad k = 1, 2, 3 \tag{2.86}$$

in a dynamical system composed of N spin-1/2 particles, where $\hat{\sigma}_k^{(n)}$ stands for the kth Pauli spin matrix of the nth particle. We easily obtain the commutation relations

$$[\hat{\Sigma}_i, \hat{\Sigma}_j] = 2i\hbar \frac{1}{N} \hat{\Sigma}_k, \quad (i, j, k) = \text{cyclic permutation of } (1, 2, 3). \tag{2.87}$$

Note that the r.h.s. of this equation vanishes in the infinite N limit. Thus the average spin components become commutable with each other in the infinite N limit, while they do not commute when N is finite. In other words, these operators tend to the *center* of the algebra as N tends to infinity. In this limit we can use the representation in which all $\hat{\Sigma}_k$s *simultaneously* become diagonal (i.e. c-numbers). Notice, however, that such a representation is not unitary equivalent to the original one for finite N, in which all $\hat{\Sigma}_k$ are uncommutable. Only when the number of degrees of freedom becomes infinite, do we obtain this remarkable jump, like in condensed matter phenomena. A nice example will come to our attention in the case of a solvable detector model in Chapter 5, in which we shall discuss how to fix the c-numbers. The appearance of an inequivalent representation is an important characteristic of systems with infinite degrees of freedom. We believe that the wave function collapse taking place in a detector is a quantum phenomenon specific to large systems with infinite degrees of freedom, as in condensed matter physics.

5   Uncertainty relations and coherent states

Consider two observables $\hat{F}$ and $\hat{G}$. Their expectation values and mean square deviations (uncertainties) in the state ψ are given by

$$\langle F \rangle \equiv \langle \psi | \hat{F} | \psi \rangle, \quad \langle G \rangle \equiv \langle \psi | \hat{G} | \psi \rangle, \quad (\Delta F)^2 \equiv \langle \psi | \hat{F}_0^2 | \psi \rangle, \quad (\Delta G)^2 \equiv \langle \psi | \hat{G}_0^2 | \psi \rangle, \tag{2.88}$$

under the normalization condition $\langle \psi | \psi \rangle = 1$, where

$$\hat{F}_0 \equiv \hat{F} - \langle F \rangle, \quad \hat{G}_0 \equiv \hat{G} - \langle G \rangle. \tag{2.89}$$

Notice that $[\hat{F}_0, \hat{G}_0] = [\hat{F}, \hat{G}]$. We can easily prove that the inequality $\langle \tilde{\psi} | \tilde{\psi} \rangle \geq 0$, valid for $|\tilde{\psi}\rangle \equiv [\xi \hat{F}_0 + i\hat{G}_0] | \psi \rangle$ and an arbitrary real number ξ, is equivalent to the relation

$$\Delta F \Delta G \geq \frac{1}{2} |\langle i[\hat{F}, \hat{G}] \rangle|. \tag{2.90}$$

This inequality yields no restriction on the value of $\Delta F \Delta G$ if the operators $\hat{F}$ and $\hat{G}$ commute ($[\hat{F}, \hat{G}] = 0$), but imposes a minimum value for this product (the uncertainty relation) if $[\hat{F}, \hat{G}] \neq 0$. The equality in (2.90) holds for the state

$$[\xi_0 \hat{F}_0 + i\hat{G}_0]|\psi_0\rangle = 0, \quad \xi_0 = -\frac{\langle i[\hat{F}, \hat{G}]\rangle_0}{2(\Delta F)^2}, \tag{2.91}$$

where $\langle \cdots \rangle_0$ stands for the expectation value in the ψ_0 state. The state ψ_0 is called the *minimal wave packet*, in the sense that the uncertainty becomes minimum in this case. The minimal wave packet is considered to describe a quantum state which is close to the corresponding classical one.

For a canonical pair $(\hat{q}_i, \hat{p}_j)$ subject to $[\hat{q}_i, \hat{p}_j] = i\hbar\delta_{ij}$, (2.90) becomes the famous uncertainty relation

$$\Delta q_i \Delta p_j \geq \frac{\hbar}{2}\delta_{ij}. \tag{2.92}$$

In this case the minimal wave packet in the q-representation is expressed by

$$\psi_0 = \frac{1}{\sqrt[4]{\prod_k 2\pi(\Delta q)_k^2}} \exp\left[\frac{i}{\hbar}\sum_k \langle p\rangle_k q_k - \sum_k \frac{(q_k - \langle q_k\rangle)^2}{4(\Delta q_k)^2}\right] \tag{2.93}$$

and we obtain the Gaussian probability distribution function

$$|\psi_0(q)|^2 = \frac{1}{\sqrt{\prod_k 2\pi(\Delta q_k)^2}} \exp\left[-\sum_k \frac{(q_k - \langle q_k\rangle)^2}{2(\Delta q_k)^2}\right]. \tag{2.94}$$

We are interested, in particular, in the case of a harmonic oscillator described by the Hamiltonian

$$\hat{H} = \sum_k \left[\frac{1}{2m_k}\hat{p}_k^2 + \frac{m_k\omega_k^2}{2}\hat{q}_k^2\right], \tag{2.95}$$

where m_k and ω_k are positive constants. In this case the uncertainties of the canonical variables in the vacuum state (i.e. the eigenstate of the Hamiltonian belonging to the lowest eigenvalue) are given by

$$(\Delta q_k)^2 = \frac{\hbar}{2m_k\omega_k}, \quad (\Delta p_k)^2 = \frac{m_k\omega_k\hbar}{2}. \tag{2.96}$$

The minimal wave packet, which is to be determined from (2.91) with the identification $\hat{F} = \hat{q}_k$ and $\hat{G} = \hat{p}_k$ for each k, is expressed as a product of states $|\alpha_k\rangle$, i.e. $|\psi_0\rangle = \prod_k |\alpha_k\rangle$, each of which obeys the eigenvalue equation

$$\hat{a}_k|\alpha_k\rangle = |\alpha_k\rangle\alpha_k, \quad \alpha_k \equiv \langle\hat{a}_k\rangle_0. \tag{2.97}$$

where $\langle\cdots\rangle_0$ is the expectation value in the $|\alpha_k\rangle$ state and

$$\hat{a}_k = \frac{1}{2(\Delta q_k)}\hat{q}_k + i\frac{1}{2(\Delta p_k)}\hat{p}_k = \sqrt{\frac{m_k\omega_k}{2\hbar}}\left(\hat{q}_k + i\frac{1}{m_k\omega_k}\hat{p}_k\right), \qquad (2.98)$$

$$\hat{a}_k^\dagger = \frac{1}{2(\Delta q_k)}\hat{q}_k - i\frac{1}{2(\Delta p_k)}\hat{p}_k = \sqrt{\frac{m_k\omega_k}{2\hbar}}\left(\hat{q}_k - i\frac{1}{m_k\omega_k}\hat{p}_k\right). \qquad (2.99)$$

These operators satisfy the canonical commutation relation

$$[\hat{a}_k, \hat{a}_{k'}^\dagger] = \delta_{kk'}. \qquad (2.100)$$

The number operator is defined by

$$\hat{n}_k \equiv \hat{a}_k^\dagger \hat{a}_k. \qquad (2.101)$$

This operator has integer eigenvalues $(0, 1, 2, \ldots)$ and satisfies

$$[\hat{n}_k, \hat{a}_k] = -\hat{a}_k, \quad [\hat{n}_k, \hat{a}_k^\dagger] = \hat{a}_k^\dagger. \qquad (2.102)$$

Because of (2.102), we are led to view $\hat{a}_k$ and $\hat{a}_k^\dagger$ as the annihilation and creation operators, respectively, which decrease and increase by one the number of *mode* k.

We usually call the eigenstate of the annihilation operator $\hat{a}_k$ the *coherent state* of mode k. Because $\hat{a}_k$ is not hermitian, its eigenvalue $\alpha_k = \langle\hat{a}_k\rangle$ (here and hereafter the subscript 0 will be suppressed for simplicity) is complex and its eigenstates $|\alpha_k\rangle$ do not constitute an orthogonal set. It is easy to find that the solution of (2.97) is given by

$$|\alpha_k\rangle = e^{-|\alpha_k|^2/2}e^{\alpha_k\hat{a}_k^\dagger}|0\rangle, \qquad (2.103)$$

under the normalization condition $\langle\alpha_k|\alpha_k\rangle = 1$, where $|0\rangle$ is the eigenstate of $\hat{n}_k$ belonging to the eigenvalue 0. This follows from the application of the relation

$$e^{-\alpha_k\hat{a}_k^\dagger}\hat{a}_k e^{\alpha_k\hat{a}_k^\dagger} = \hat{a}_k + \alpha_k. \qquad (2.104)$$

From (2.97) we know that $\langle\alpha_k|$ is a left-eigenstate of $\hat{a}_k^\dagger$ belonging to α_k^*

$$\langle\alpha_k|\hat{a}_k^\dagger = \alpha_k^*\langle\alpha_k|. \qquad (2.105)$$

Equations (2.97) and (2.105) mean that the expectation value of the number operator and its mean square (uncertainty) in the coherent state $|\alpha_k\rangle$ are given by

$$\langle n_k\rangle = \langle\alpha_k|\hat{a}_k^\dagger\hat{a}_k|\alpha_k\rangle = |\alpha_k|^2, \quad (\Delta n_k)^2 = \langle n_k\rangle. \qquad (2.106)$$

Note that this uncertainty Δn_k comes from the so-called *zero-point* fluctuation, which is nothing but the *vacuum fluctuation* in the case of a laser field, as described in terms of the electric field at the end of this section. In the infinite $\langle n_k \rangle$ limit, the relative uncertainty becomes vanishingly small, because

$$\frac{(\Delta n_k)^2}{\langle n_k \rangle^2} = \frac{1}{\langle n_k \rangle} \to 0. \tag{2.107}$$

In this context, the coherent state of the field is considered to be very close to a classical field.

By making use of the well-known formula

$$e^{\hat{A}} e^{\hat{B}} = e^{\hat{A}+\hat{B}+[\hat{A},\hat{B}]/2} \tag{2.108}$$

for operators $\hat{A}$ and $\hat{B}$, both of which commute with $[\hat{A}, \hat{B}]$, we can rewrite (2.103) as

$$|\alpha_k\rangle = \hat{D}_k|0\rangle, \quad \hat{D}_k \equiv e^{\alpha_k \hat{a}_k^\dagger - \alpha_k^* \hat{a}_k}, \tag{2.109}$$

where $\hat{D}_k$ is a displacement operator yielding

$$\hat{D}_k^\dagger \hat{a}_k \hat{D}_k = \hat{a}_k + \alpha_k, \quad \hat{D}_k^\dagger \hat{a}_k^\dagger \hat{D}_k = \hat{a}_k^\dagger + \alpha_k^*. \tag{2.110}$$

For the number states defined by

$$\hat{n}_k|n_k\rangle = n_k|n_k\rangle, \tag{2.111}$$

the annihilation (creation) operator decreases (increases) the number n_k by one

$$\hat{a}_k|n_k\rangle = \sqrt{n_k}|n_k - 1\rangle, \quad \hat{a}_k^\dagger|n_k\rangle = \sqrt{n_k + 1}|n_k + 1\rangle. \tag{2.112}$$

By using these formulae, we obtain

$$\langle n_k|\alpha_k\rangle = \frac{(\alpha_k)^{n_k}}{\sqrt{n_k!}} e^{-|\alpha_k|^2/2}, \tag{2.113}$$

so that the probability of finding the number state $|n_k\rangle$ in the coherent state $|\alpha_k\rangle$ is given by the Poisson distribution

$$|\langle n_k|\alpha_k\rangle|^2 = \frac{|\alpha_k|^{2n_k}}{n_k!} e^{-|\alpha_k|^2}. \tag{2.114}$$

This is a remarkable feature of the coherent state. Equation (2.113) also gives the following expansion formula

$$|\alpha_k\rangle = e^{-|\alpha_k|^2/2} \sum_{n_k=0}^{\infty} \frac{(\alpha_k)^{n_k}}{\sqrt{n_k!}} |n_k\rangle, \tag{2.115}$$

which is consistent with (2.103) and (2.109).

Suppose that we have two coherent states

$$|\alpha_k\rangle = e^{-|\alpha_k|^2/2}e^{\alpha_k \hat{a}_k^\dagger}|0\rangle, \quad |\beta_\ell\rangle = e^{-|\beta_\ell|^2/2}e^{\beta_\ell \hat{a}_\ell^\dagger}|0\rangle. \tag{2.116}$$

We easily see that their inner product has modulus

$$|\langle \beta_\ell | \alpha_k \rangle| = e^{-(|\alpha_k|^2+|\beta_\ell|^2)/2}|e^{\beta_\ell^* \alpha_k}| = e^{-|\alpha_k - \beta_\ell|^2/2}. \tag{2.117}$$

Therefore, two coherent states are approximately orthogonal if their eigenvalues are very different from each other. Furthermore, it is easy to show the normalization condition and the completeness relation for $\{|\alpha_k\rangle\}$

$$\text{Tr}|\alpha_k\rangle\langle\alpha_k| = 1, \quad \frac{1}{\pi}\int |\alpha_k\rangle\langle\alpha_k| d^2\alpha_k = 1. \tag{2.118}$$

In quantum optics[116,197,119] we often use coherent states in order to represent a strong laser beam. Consider a plane electromagnetic wave of mode μ (having wave number k and polarization r) propagating along the z axis, described by the electric field

$$\hat{E}_\mu = i\sqrt{\frac{\hbar\omega_k}{2\Omega}}[\hat{a}_\mu e^{i(kz-\omega_k t)} - \hat{a}_\mu^\dagger e^{-i(kz-\omega_k t)}], \tag{2.119}$$

where ω_k and Ω are frequency and volume, respectively. Note that we have used the box-normalization in (2.119). If we apply the above operators to the coherent state of mode k we obtain a useful formula to describe a strong laser beam. As was mentioned above, the vacuum fluctuation plays an important role in the dephasing process.

Chapter 3

CRITICAL REVIEW OF MEASUREMENT THEORIES

In Chapter 1, we have seen that the wave function collapse by measurement is described as a dephasing process between the branch waves, caused by a measuring apparatus or a detector. However, many other ideas and viewpoints have been proposed under the same terminology of "wave function collapse," and many discussions have taken place among several schools of thoughts. By examining the notion of wave function collapse, in this chapter, we present a critical review of a few famous measurement theories and some important debates and paradoxes. We have already outlined some of them in Chapter 1. Here we supplement that discussion with some technical details.

1 Collapse of the wave function

In this section we shall first reexamine the notion of wave function collapse and then formulate it in a proper way. Remember that we have already ruled out the so-called naive WFC for the three reasons mentioned in Section 1.2 of Chapter 1.

Consider the measurement of an observable $\hat{F}$ of the object system Q in a superposed state

$$|\psi_{\mathrm{I}}^{Q}\rangle = \sum_{i} c_i |\chi_i\rangle, \quad c_i = \langle \chi_i | \psi_{\mathrm{I}}^{Q}\rangle. \tag{3.1}$$

Assume that $\hat{F}$ is a linear self-adjoint operator of system Q, having real eigenvalues and corresponding eigenfunctions. In some cases we can choose $|\chi_i\rangle$ to be an eigenstate of the observable $\hat{F}$ belonging to a real eigenvalue λ_i.

As for the third criticism against the naive WFC (i.e. the lack of probabilistic rule (1.9) in it), von Neumann[195] himself improved the description of the WFC in the following way

$$\hat{\rho}_{\mathrm{I}}^{Q} = |\psi_{\mathrm{I}}^{Q}\rangle\langle\psi_{\mathrm{I}}^{Q}| = \sum_{i}\sum_{j} c_i c_j^* |\chi_i\rangle\langle\chi_j| \;\rightarrow\; \hat{\bar{\rho}}_{\mathrm{F}}^{Q} = \sum_{k} |c_k|^2 \hat{\xi}_k^{Q}, \tag{3.2}$$

where $\hat{\xi}_k^Q \equiv |\chi_k\rangle\langle\chi_k|$ is the projection operator onto the kth eigenstate, and the subscripts I and F refer to the initial and final states, respectively. The bar is used for the (completely) mixed state with no phase correlation among the branch waves [see (2.69)]. This expression describes well a process in which all the phase correlations among different eigenstates are erased, so that we obtain a sum of exclusive probabilities of finding each eigenstate, along the line of thought roughly sketched in Section *1.2* of Chapter 1.

However, (3.2) is still not satisfactory, because in some cases this description leads to contradictions. Consider, for example, the case $c_1 = c_2 = 1/\sqrt{2}$, corresponding to a dichotomic observable $\hat{F}$. In this case, (3.2) becomes

$$\hat{\rho}_I^Q \rightarrow \bar{\hat{\rho}}_F^Q = \frac{1}{2}(\hat{\xi}_1^Q + \hat{\xi}_2^Q). \tag{3.3}$$

On the other hand, by defining $\chi_\pm = (\chi_1 \pm \chi_2)/\sqrt{2}$, we easily see that (3.3) turns into

$$\hat{\rho}_I^Q \rightarrow \bar{\hat{\rho}'}_F^Q = \frac{1}{2}(\hat{\xi}_+^Q + \hat{\xi}_-^Q) = \bar{\hat{\rho}}_F^Q, \tag{3.4}$$

where $\hat{\xi}_\pm^Q \equiv |\chi_\pm\rangle\langle\chi_\pm|$. Note that $\bar{\hat{\rho}}_F^Q = \bar{\hat{\rho}'}_F^Q$ is an identity, and observe that (3.3) describes a measurement of the observable $\hat{F}$, while (3.4) describes a measurement of another observable $\hat{G}$ with eigenstates $\chi_\pm$. In general, $[\hat{F}, \hat{G}] \neq 0$, so that one is led to the contradiction that (3.2) describes the incompatible measurements of two uncommutable observables at the same time. This contradiction was pointed out to us by Watanabe,[198] and is deeply rooted into the well-known phenomenon of nonunique decomposability of mixed states,[47,34] which can also be used to point out that some (statistical) interpretations concerning the behaviour of EPR-correlated particles[53] lead to inconsistencies.

In order to circumvent this contradiction, we must modify the expression (3.2) for the wave function collapse, by introducing the states of the detector system (denoted by D) in the following way

$$\hat{\Xi}_I^{tot} = \hat{\rho}_I^Q \otimes \hat{\sigma}_I^D = \sum_i \sum_j c_i c_j^* |\chi_i\rangle\langle\chi_j| \otimes \hat{\sigma}_I^D$$

$$\longrightarrow \hat{\Xi}_{Ft}^{tot} = \sum_k |c_k|^2 \hat{\xi}_{F(k)t}^Q \otimes \hat{\sigma}_{F(k)t}^D, \tag{3.5}$$

where $\hat{\Xi}^{tot}$ is the density matrix of the total system Q+D, $\hat{\sigma}_I^D$ stands for the initial density matrix of D, $\hat{\xi}_{F(k)t}^Q$ for the final Q state corresponding to the kth eigenstate, and $\hat{\sigma}_{F(k)t}^D$ for the final density matrix of D, displaying the kth eigenvalue of $\hat{F}$. The subscript t (time variable) in the far r.h.s. denotes the free temporal evolution of systems Q and D without any interaction between them. In particular, $\hat{\sigma}_{F(k)t}^D$ describes secondary processes such as counter discharge phenomena. The above inconsistency pointed out by Watanabe does not occur anymore, owing to the presence of the D states. This is the measurement process which we are going to explain in this book.

Incidentally, since the states of D *must* be explicitly included into the description of quantum measurements in order to avoid the afore-mentioned inconsistency, one may say that quantum mechanics, unlike classical physics, is not consistent with *naive realism*. Indeed, we need not explicitly introduce any detector variables or states in order to discuss the measurement problem in classical physics.

Observe that the wave function collapse by measurement must be written as a transition process of the total density matrix from $\hat{\Xi}_{\mathrm{I}}^{\mathrm{tot}}$, which includes off-diagonal components, to $\hat{\Xi}_{\mathrm{Ft}}^{\mathrm{tot}}$, which has no off-diagonal part. Henceforth, we shall use the expressions "diagonal" and "off-diagonal" parts of the density matrix, referring only to the eigenstates χ_i of system Q. We stress that (3.5) is the final goal of a quantum theory of measurements. The r.h.s. of (3.5) corresponds to a set of mutually exclusive probabilistic events. Notice again that (3.5) describes the fact that only one event, out of all the mutually exclusive probabilistic events, will eventually take place, and that once one event happens, all others never occur. This is typical of probabilistic phenomena, common to both classical and quantum mechanics. Remember the discussion in Sections *1.1* and *1.2* of Chapter 1.

We are therefore led to the exact formulation of the measurement problem in the following way: Can we derive the process (3.5) by applying quantum mechanics to the total system Q+D? This problem will be solved in Chapter 4.

2 Von Neumann-Wigner theory and related topics

We have already outlined the von Neumann-Wigner theory in Section *2.1* of Chapter 1. In this section we rediscuss this theory, together with some technical details, because this approach is worth discussing for many reasons.

2.1 The von Neumann-Wigner approach

We can regard the measurement process of D on Q as a collision process between Q and a local system of D (denote it by the same letter), as we have seen in Chapter 1. Denote the total Hamiltonian by $\hat{H}$, the free Hamiltonian by $\hat{H}_0 = \hat{H}^{\mathrm{Q}} + \hat{H}^{\mathrm{D}}$ (the free Hamiltonians of the Q and D systems, respectively), and the interaction Hamiltonian by $\hat{H}_{\mathrm{QD}}$. For the measurement of an observable $\hat{F}$ of Q in a superposed state $|\psi_{\mathrm{I}}^{\mathrm{Q}}\rangle$ as given by (3.1), this collision (or scattering) process is written in terms of the S-matrix in the following way:

$$e^{-i\hat{H}t/\hbar}\{|\chi_i\rangle \otimes |\Phi_0^{\mathrm{D}}\rangle\} \stackrel{t\to\infty}{\longrightarrow} e^{-i\hat{H}_0t/\hbar}\hat{S}\{|\chi_i\rangle \otimes |\Phi_0^{\mathrm{D}}\rangle\}, \tag{3.6}$$

where $|\Phi_0^D\rangle$ stands for the initial detector state. This is nothing but the definition of the S-matrix in the present case: see (2.57) and the discussion in Section 2.2, Chapter 2.

One of the simplest ways of modelling a detector is exemplified by the following action of the S-matrix

$$\hat{S}\{|\chi_i\rangle \otimes |\Phi_0^D\rangle\} = |\chi_i\rangle \otimes |\Phi_i^D\rangle, \qquad (3.7)$$

where $|\Phi_i^D\rangle$ is the final detector state. In other words, this detector model represents an instrument that changes its state from $|\Phi_0^D\rangle$ to $|\Phi_i^D\rangle$, according to the object eigenstate $|\chi_i\rangle$. Surely, this is one of the simplest possible detector models.

Von Neumann and Wigner strongly asserted that quantum mechanics, in particular the superposition principle, should be strictly applied to the measurement process. With the help of (3.6) and (3.7), therefore, they obtained

$$|\Psi_I^{tot}\rangle \equiv |\psi_I^Q\rangle \otimes |\Phi_0^D\rangle = \sum_i c_i |\chi_i\rangle \otimes |\Phi_0^D\rangle \longrightarrow \sum_i c_i |\chi_i\rangle \otimes |\Phi_i^D\rangle \equiv |\tilde{\Psi}_F^{tot}\rangle \quad (3.8)$$

for the whole measurement process. This is often called the von Neumann measurement process.

However, the phase correlation among different Q eigenstates is still present in $|\tilde{\Psi}_F^{tot}\rangle$, as we easily understand by writing the corresponding density matrix

$$\hat{\tilde{\rho}}_F^{tot} = |\tilde{\Psi}_F^{tot}\rangle\langle\tilde{\Psi}_F^{tot}| = \sum_k |c_k|^2 \hat{\xi}_k^Q \otimes |\Phi_k^D\rangle\langle\Phi_k^D| + \sum_i \sum_{j \neq i} c_i c_j^* |\chi_i\rangle\langle\chi_j| \otimes |\Phi_i^D\rangle\langle\Phi_j^D|,$$
$$(3.9)$$

where $\hat{\xi}_k^Q = |\chi_k\rangle\langle\chi_k|$. Obviously, the above is still a projection operator representing a pure state with non-vanishing off-diagonal components (the last term in the r.h.s.). Thus, we are led to the conclusion that the von Neumann-Wigner approach can never realize the wave function collapse as a physical process.

As is well known and was already mentioned in Chapter 1, the von Neumann-Wigner theory required that the measuring process is not completed by the interaction of Q with D, which is instead followed by the so-called von Neumann's chain of measurements, connecting Q to the observer via many steps. Eventually the theory brings in an "Abstraktes Ich" or "consciousness" at the end of the chain, which should be responsible for the wave function collapse by measurement. On the basis of these arguments, Wigner claimed that quantum mechanics is incomplete. This theory provoked the famous paradoxes of Schrödinger's cat and Wigner's friend, roughly sketched in Section 2.1 of Chapter 1. The whole state in the cat paradox is considered to be

$$|\Psi\rangle = \frac{1}{\sqrt{2}} \left[|\text{alive}\rangle_{cat} \otimes |+\rangle_Q + |\text{dead}\rangle_{cat} \otimes |-\rangle_Q\right], \qquad (3.10)$$

where the subscripts "cat" and Q stand for the cat and Q states, and $\pm$ for the excited and ground states of Q, respectively. The r.h.s. of this expression is to be compared with $|\tilde{\Psi}_F^{tot}\rangle$ of (3.8), so that we cannot definitely say whether the cat is alive or dead. In Chapter 4, we shall discuss a possible experimental realization of the Schrödinger cat, from the mesoscopic point of view, in close connection with the many-Hilbert-space theory.

We should also remark that the detector states $|\Phi_i^D\rangle$ in (3.8) are not always subject to the orthogonality condition

$$\langle \Phi_i^D | \Phi_j^D \rangle = \delta_{ij}. \qquad (3.11)$$

Originally, we assumed that $\{|\Phi_i^D\rangle\}$ is a set of eigenstates of a macroscopic observable of D, like a pointer position, so that the von Neumann process takes place within the orthogonality condition of a self-adjoint observable. However, Araki and Yanase[8,209] showed that a measurement process of an observable which does not commute with a conserved (additive) quantity cannot be described by (3.8) with the orthogonality condition (3.11). In spite of this, they also proved that an approximate measurement of such an observable is possible in the original sense of a von Neumann process, up to any desired accuracy, if we use an adequate measuring apparatus endowed with a huge number of degrees of freedom. From this point of view, it is possible to find a measuring apparatus that yields a measurement process of the von Neumann type, for which the following asymptotic orthogonality holds:

$$\langle \Phi_i^D | \Phi_j^D \rangle = \delta_{ij} + (1 - \delta_{ij})O(\epsilon), \quad \epsilon \xrightarrow{N \to \infty} 0, \qquad (3.12)$$

where N is a parameter corresponding to the number of degrees of freedom of the apparatus system.

Conventionally, we can accept (3.8) as a typical spectral decomposition process (1.2). For example, recall (1.13) in the Stern-Gerlach experiment (see Fig. 1.4), and make the following correspondence

$$\psi_0 \leftrightarrow \Psi_I^{tot}, \quad \phi_{a/b} \leftrightarrow \Phi_i^D, \quad \phi \leftrightarrow \Phi_0^D,$$
$$\psi_1 \leftrightarrow \Psi_F^{tot}, \quad u_{a/b} \leftrightarrow \chi_i. \qquad (3.13)$$

However, this correspondence sometimes leads to the misleading conclusion[90] that (1.13), namely the simple passage through the magnetic field (without detection), can yield a genuine quantum measurement. Equation (1.13) is, of course, only a spectral decomposition, but does not give dephasing and does not lead to the WFC. As was repeatedly emphasized, a quantum measurement is completed by two successive steps: spectral decomposition and detection. Remember that only the latter is responsible for the dephasing.

The ergodic amplification theory, which was one of the main antagonists of the von Neumann-Wigner theory, was reviewed and criticized in Chapter 1.

2.2 Wave function collapse based on partial tracing or partial inner product and environment theory

In spite of the above arguments, there are still some authors who believe that the von Neumann process (3.8), without any additional manipulation, can describe the wave function collapse, provided that the D states satisfy the orthogonality condition (3.11) or (3.12). Such a derivation of the WFC from (3.8) is based on the computation of the partial trace or the partial inner product with respect to the D states.

By computing the trace of the density matrix $\hat{\rho}_F^{\text{tot}}$ in (3.9) [corresponding to the von Neumann process (3.8)] with respect to the D states, we immediately obtain

$$\text{Tr}_D \hat{\rho}_F^{\text{tot}} = \sum_k |c_k|^2 \hat{\xi}_k^Q \tag{3.14}$$

if we make use of the orthogonality condition (3.11) or (3.12). The r.h.s. is nothing but (3.2), so that some authors believe that the WFC can be derived. Against this kind of argument, however, we recall our discussion in Section 1 and our conclusion that the process leading to the r.h.s. of (3.14) is unsatisfactory.

In order to make our point clearer, let us explicitly show that we cannot identify the wave function collapse with a von Neumann process plus the orthogonality of the D states. To this end, first decompose the final state density matrix $\hat{\rho}_F^{\text{tot}}$ into the sum of its diagonal and off-diagonal parts

$$\hat{\rho}_F^{\text{tot}} = \hat{\rho}_{\text{diag}} + \hat{\rho}_{\text{off}},$$
$$\hat{\rho}_{\text{diag}} = \sum_k |c_k|^2 \hat{\xi}_k^Q \otimes |\Phi_k^D\rangle\langle\Phi_k^D|, \quad \hat{\rho}_{\text{off}} = \sum_i \sum_{j\neq i} c_i c_j^* |\chi_i\rangle\langle\chi_j| \otimes |\Phi_i^D\rangle\langle\Phi_j^D|, \tag{3.15}$$

where we have suppressed the subscript F for simplicity. By making use of the relation $(\hat{\rho}_F^{\text{tot}})^2 = \hat{\rho}_F^{\text{tot}}$, we easily compute

$$(\hat{\rho}_{\text{diag}})^2 - \hat{\rho}_{\text{diag}} = \hat{\rho}_{\text{off}} - (\hat{\rho}_{\text{off}})^2 - (\hat{\rho}_{\text{diag}}\hat{\rho}_{\text{off}} + \hat{\rho}_{\text{off}}\hat{\rho}_{\text{diag}}). \tag{3.16}$$

By calculating the partial trace with respect to the D states before taking the limit for $N \to \infty$, we obtain

$$\text{Tr}_D \hat{\rho}_{\text{off}} = O(\epsilon), \tag{3.17}$$
$$\text{Tr}_D(\hat{\rho}_{\text{off}})^2 = -\text{Tr}_D\left[(\hat{\rho}_{\text{diag}})^2 - \hat{\rho}_{\text{diag}}\right] + O(\epsilon)$$
$$= \sum_k |c_k|^2(1 - |c_k|^2)\hat{\xi}_k^Q + O(\epsilon). \tag{3.18}$$

This means that even though its trace vanishes, $\hat{\tilde{\rho}}_{\text{off}}$ itself does not vanish, even in the infinite N limit, because $\text{Tr}_{\text{D}}(\hat{\tilde{\rho}}_{\text{off}})^2 \neq 0$. Therefore, this kind of approach can never yield the exact wave function collapse, as formalized in the preceding subsection. We stress that the WFC is not derived unless the disappearance of the off-diagonal components is explicitly shown.

In order to shed light on the connection between the role of the partial inner product and the orthogonality condition (3.11) and the difference with a genuine WFC, let us first consider the spectral decomposition (1.13), for which the orthogonality among the branch waves is always implicitly assumed. In this case, the supports of the two branch waves running in channels A and B do not overlap and the respective spin states are orthogonal, so that we can never observe interference between them, by a detector D_0 placed in the final channel 0, even if both waves are recombined in the final channel. Even in this case, however, the two waves do not lose coherence and can display a beautiful interference pattern, which can be brought to light by making use of a sophisticated technique, as was shown by many experiments of neutron interference.[1,11,161] At this point, we have to mention the remarkable neutron interference experiments performed by Werner's and Rauch's groups,[40,93,87,162] evidencing that the apparent disappearance of interference does not always mean loss of coherence. Remember that no coherence does imply no interference, but no interference does not necessarily signify no coherence. In this context, we stress that we should not be satisfied by the absence of interference between the branch waves; rather, we have to explicitly show the absence of coherence between them, if we want to derive the WFC in our measurement theory. For this reason, we cannot regard the spectral decomposition as a WFC process. Neutron interferometry experiments, which have played and still play a very important role in fundamental quantum mechanical problems, will be reviewed in Chapter 6.

In order to analyze the partial inner product and the orthogonality condition (3.11) we shall consider two examples[177,181,143] of possible detector models. Both models are based on the technique of computing the partial inner product and on the concept of "environment."

Let us start by briefly reviewing the main points of the analysis performed by Scully and Walther (SW),[177] who consider a polarized neutron interacting with two micromasers. The ith micromaser $(i = 1, 2)$, whose initial state vector is $|\Phi_i^0\rangle$, is placed along the ith arm of an interferometer. The initial neutron + micromasers state vector is

$$\Psi(t = 0) = [\psi_1(r, 0) + \psi_2(r, 0)]|\uparrow\rangle \otimes |\Phi_1^0 \Phi_2^0\rangle, \qquad (3.19)$$

where $\psi_i(r, t)$ is the neutron wave packet running through route i, and the initial neutron spin is assumed to be "up." If the micromasers are prepared in a state that can provoke   spin flip on the neutron with probability close to one, the final state vector is

$$\Psi(t) = \psi_1(r,t)|\downarrow\rangle \otimes |\Phi_1^F \Phi_2^0\rangle + \psi_2(r,t)|\downarrow\rangle \otimes |\Phi_1^0 \Phi_2^F\rangle, \qquad (3.20)$$

where $|\Phi_i^F\rangle$ denotes the state of the micromaser after interaction. The interference term is

$$2\mathrm{Re}\psi_1^* \psi_2 \langle \Phi_1^F \Phi_2^0 | \Phi_1^0 \Phi_2^F \rangle, \qquad (3.21)$$

and one can see that the presence of the masers provokes a reduction of the interference term, which turns out to be multiplied by the factor

$$\langle \Phi_1^F \Phi_2^0 | \Phi_1^0 \Phi_2^F \rangle. \qquad (3.22)$$

SW consider two interesting cases. If the micromaser cavity is prepared in a coherent state

$$|\Phi_i^0\rangle = |\alpha_i\rangle, \qquad i = 1,2 \qquad (3.23)$$

one can write, to a good approximation,

$$|\Phi_i^F\rangle \simeq |\alpha_i\rangle, \qquad i = 1,2 \qquad (3.24)$$

because the "classical" coherent field is not changed much by the addition of a single photon associated with the neutron spin flip. In this way

$$\langle \Phi_1^F \Phi_2^0 | \Phi_1^0 \Phi_2^F \rangle \simeq \langle \alpha_1 \alpha_2 | \alpha_1 \alpha_2 \rangle = 1, \qquad (3.25)$$

and interference is not reduced, even though spin flipping has occurred. On the other hand, if the micromaser cavity is initially in a number state

$$|\Phi_i^0\rangle = |n_i\rangle, \qquad i = 1,2 \qquad (3.26)$$

after the neutron spin flip we have

$$|\Phi_i^F\rangle \simeq |n_i + 1\rangle, \qquad i = 1,2 \qquad (3.27)$$

and interference disappears due to the orthogonality

$$\langle \Phi_1^F \Phi_2^0 | \Phi_1^0 \Phi_2^F \rangle \simeq \langle n_1 + 1, n_2 | n_1, n_2 + 1 \rangle = 0. \qquad (3.28)$$

SW argue that this provides a counterexample to Bohr and Heisenberg "random-phase" argument, according to which interference is lost in a double-slit experiment when the interaction between the Q particle and the experimental apparatus yields a random phase shift between the two branch waves of the former. According to SW, the loss of interference due to the vanishing of the scalar product in (3.28) reflects a loss of coherence between the two branch waves of the particle.

Even though we agree with SW's general discussion, we cannot accept their last conclusion: The vanishing of the interference term does *not imply at all* a loss of coherence. A loss of coherence corresponds to the WFC and this is *not* what happens in the example proposed by SW. This can be easily shown by writing the final density matrix for the neutron + micromasers system

$$\rho_t = |\psi_1|^2 \otimes |\Phi_1^F \Phi_2^0\rangle\langle\Phi_1^F \Phi_2^0| + |\psi_2|^2 \otimes |\Phi_1^0 \Phi_2^F\rangle\langle\Phi_1^0 \Phi_2^F|$$
$$+\psi_1^* \psi_2 |\Phi_1^0 \Phi_2^F\rangle\langle\Phi_1^F \Phi_2^0| + \text{h.c.}, \tag{3.29}$$

where the common term $|\downarrow\rangle\langle\downarrow|$ has been suppressed for simplicity. One sees immediately that the off-diagonal part of the density matrix does not vanish, as a result of the interaction. This conclusion is true *independently* of the final state of the maser cavity, which can be either (3.24) or (3.27). In this sense, we can state that the process analyzed by SW corresponds to a loss of interference, and not to the WFC. Indeed, in (3.29), the quantum mechanical coherence is fully kept between the two branch waves of the neutron + micromasers system.

It is worth mentioning that the above-mentioned idea has led to very interesting experiments on cavity QED.[71] Such experiments are also relevant for the practical realization of Schrödinger's "cat" states in the mesoscopic domain.

Let us turn now our attention to a model studied by Stern, Aharonov and Imry (SAI).[181] They consider an Aharonov-Bohm interference experiment of an electron on a ring. The electron, whose coordinate is x, is split into two branch waves $\ell(x)$ and $r(x)$, crossing the ring along the right and left arm, respectively. The initial wave function of the electron + ring system is written as

$$\Psi(t = 0) = [\ell(x) + r(x)] \otimes \chi_0(\eta), \tag{3.30}$$

where η denotes the set of coordinates describing the state of the ring and χ_0 its initial wave function. At time τ_0, when interference is observed, the wave function is

$$\Psi(\tau_0) = \ell(x, \tau_0) \otimes \chi_\ell(\eta) + r(x, \tau_0) \otimes \chi_r(\eta), \tag{3.31}$$

where $\chi_{\ell,r}(\eta)$ are the states of the ring after interaction. The interference term is given by

$$2\,\text{Re}\left(\ell^*(x, \tau_0)r(x, \tau_0)\int d\eta\,\chi_\ell^*(\eta)\chi_r(\eta)\right), \tag{3.32}$$

and it is reduced by the factor

$$\int d\eta\chi_\ell^*(\eta)\chi_r(\eta). \tag{3.33}$$

Notice that, since the state of the ring is not observed, its coordinates η are integrated over. Mathematically, this corresponds to taking the scalar product between the two final states of the ring, exactly as in the example proposed by Scully and Walther.

Observe that also in this case the final density matrix of the electron + ring system is

$$\langle x, \eta | \rho_{\tau_0} | x, \eta \rangle = |\ell(x, \tau_0)|^2 \otimes |\chi_\ell(\eta)|^2 + |r(x, \tau_0)|^2 \otimes |\chi_r(\eta)|^2$$
$$+ r(x, \tau_0)\ell^*(x, \tau_0)\chi_r(\eta)\chi_\ell^*(\eta) + \text{c.c.}, \qquad (3.34)$$

so that the off-diagonal part does *not* vanish, as a result of the interaction. As already pointed out in connection with the previous example, this process describes a (partial) loss of interference, and *not* the wave function collapse.

Unlike Scully and Walther, SAI argue that this loss of interference can alternatively be ascribed to a dephasing of the two electron states. But even in this case, it is not clear why one has to take the partial inner product or the partial trace with respect to the apparatus states. We stress again that we have to show the *vanishing of the off-diagonal components* of the total density matrix, without resorting to the partial inner product or the partial trace. In the above example, the ring plays the role of "environment." This leads us to discuss partial tracing in connection with the so-called environment theories.

The orthogonal decomposition (3.8) represents the theoretical base for the *environment* theory.[210,207,91,211] The advocates of this theory believe that in a measurement process the environment around the object system, including the measuring apparata, yields a decomposition of the type (3.8) and then provokes the WFC (3.2) with the help of (3.11) or (3.12). As can be seen in Ref. 211, there are some physical reasons for this process to take place. However, we do not agree with such an approach, for all the theoretical reasons explained above. It is worth stressing that the properties (3.11) and (3.12) require a sort of "astuteness" of the environment, almost as if it "knew" in advance that it is supposed to yield a quantum measurement.

We already discussed the status of von Neumann's projection postulate in quantum mechanics. The same conclusions are true for the technique of partial tracing over the states of the environment. The two procedures are equivalent. The technique of tracing over the states of the environment, as well as the projection postulate, are very useful from a computational point of view. On the other hand, they *do not correspond to any physical operation*. We refuse to regard sentences like "We do not look at the states of the environment" and/or "We neglect the unobserved degrees of freedom" on the same footing as physical operations, because this would make the measurement theory depend upon the human act of observation.

Notice that partial tracing is just a trick (although a very convenient one!): the evolution of the total system (Q+D) is always unitary, for each elementary process, but at a certain moment the (theoretical) physicist decides to trace over what he or she does not observe, obtaining a nonunitary evolution. However, the total quantum system (that does not know that the theoretical physicist has computed the partial trace) is still evolving in a unitary way. One is led to wonder about what would happen if another theoretical physicist would decide *not* to compute the partial trace. It is hard for us to believe that Nature behaves according to what theoretical physicists decide to compute, and we think that partial tracing can only be regarded as a convenient mathematical tool.

There are other important arguments against the so-called environment theories. One of the strongest, in our opinion, was put forward by Venugopalan, Kumar and Ghosh.[192] These three authors pointed out that environment theories (Zurek's in particular) hinge upon the concept of "preferred basis," into which the density matrix of the object system diagonalizes after interacting with the environment. In all the examples considered,[207,211] the coupling of the Q particle with the environment occurs via the position operator of the former, so that the position basis naturally emerges as the preferred Q basis. However, an *exact* solution of the equations of motion[192] shows that the Q system becomes diagonal in the *momentum* (rather than position) basis. This shows, in our opinion, that the very concept of "preferred basis" is obscure and poorly defined. It is hard for us to believe that such a fundamental issue as the quantum measurement problem can be solved in this way. We believe instead that a clear-cut solution is required, based on well-defined concepts and ideas.

In spite of the above-mentioned strong criticisms, however, we might say that in a loose sense, the many-Hilbert-space theory can be considered close to the environment theory, because one may regard the matter constituting the detector (and endowed with the MHS structure) as a sort of "environment." However, there is an essential difference between the MHS theory and all environment theories, because the former never relies upon the technique of partial tracing. This difference is, so to speak, a matter of attitude toward physics. We also feel that this difference, as well as the philosophical attitudes underlining the two approaches, is profound.

Like the environment theory, also, the many-worlds interpretation[54,202] outlines a theory of measurement within the framework of the orthogonal decomposition (3.8). As we saw in Chapter 1, in this interpretation each branch state, specified by a different measured value, together with the measuring apparatus, branches into its own world. We shall not repeat the criticisms put forward in Chapter 1.

2.3   An experimental test by a double Stern-Gerlach apparatus

In order to counter the idea that the orthogonal decomposition suffices to provoke the WFC and yield a complete quantum measurement, we propose an experimental test to examine whether the orthogonal decomposition can provoke dephasing. This experiment may be called the *double Stern-Gerlach experiment*, and is carried out by observing neutron interference in the following way. See Fig. 3.1, in which a bar stands for a neutron semi-transparent mirror and a bar in a box for a magnetic mirror. The magnetic mirror enables us to perform the spectral decomposition, i.e. it divides a neutron beam with indefinite spin into two branch waves with definite spins, as is easily understood by looking at Fig. 3.1. An incident beam is divided by the first semi-transparent mirror into two branch waves running into the upper and lower channels, irrespectively of their spin. The two branch waves are then (spin) spectrally decomposed by the magnetic mirrors. Finally, the two branch waves relative to the same spin direction, one originating from the upper magnetic mirror and the other from the lower one, are recombined by the second semi-transparent mirror and finally forwarded to the final channel towards a neutron detector (denoted by a circle).

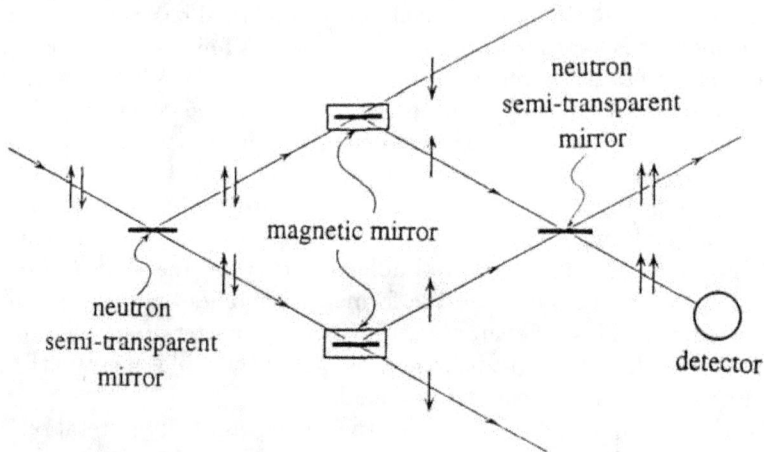

Fig. 3.1. Double Stern-Gerlach experiment.

The point is whether the waves going to the final channel keep their coherence or not. According to those measurement theories which predict WFC in this case, we cannot observe interference by the detector placed in the final channel, because the WFC already took place during the passage through the magnetic mirrors and consequently the two waves lost their coherence. Even in the many-worlds interpretation and the environment theory, the situation is similar to this case if they managed to derive decoherence with a direct link to the orthogonal decomposition. On the contrary, the MHS theory clearly predicts the appearance of interference because the separation by the magnetic mirror is only a spectral decomposition preserving coherence. Many experimentalists naturally expected that interference would be observed in this experiment. Indeed a Japanese neutron-physics group has experimentally confirmed this expectation.[52]

3 Jauch's approach and the spontaneous localization model

Let us conclude this critical review of measurement theories by illustrating a few more approaches. A tentative way out of the quantum measurement problem was put forward some years ago by Jauch.[89] His proposal is to require a "classical" property for some macroscopic apparata: The possible observables of a macrosystem are restricted to a commutable set, so that for such a "classical" system the physical states fall into equivalence classes which cannot be distinguished by any observation performed on the system. In particular, pure and mixed states turn out, under appropriate conditions, to belong to the *same* equivalence class, and are therefore completely equivalent from a physical point of view. Such states are defined as "classical," in the sense mentioned before. This obviously "solves" the measurement problem, because the states Ψ_F, in Eq. (3.8) and $\widehat{\Xi}_{Ft}^{tot}$ in Eq. (3.5) belong to the same equivalence class, and are therefore physically indistinguishable.

It must be noted, however, that the conceptual difficulties inherent to quantum measurements are not resolved. Indeed, the very problem of the loss of quantum mechanical coherence is just avoided altogether: By defining an observable as "classical," one does not give a clear prescription that enables us to pick out "classical" object. This is true, in particular, for the detector observables, which must be endowed with such a classical property by definition. Even more important, the *classical behaviour* of a certain system must be *proven* to be a good approximation, and is not to be simply postulated from the outset.

In this sense, the "solution" proposed by Jauch does not *solve* the measurement problem. It simply recasts it in a different form, by shifting it to a different level of description. Nevertheless, it would be superficial to state that the above-mentioned approach does not yield anything new. Its formal elegance suggests that it must be possible to define observables for macroscopic systems that approximate a classical behaviour in some limit. Examples of this kind will appear even in the framework of the MHS theory, as will be seen in the next chapter. Interesting steps forward in this direction have also been made in the context of the emergence of equivalence classes for macroscopic systems.[208,95] It seems therefore that Jauch's approach might be used at least as a guideline for the description of measuring processes.

An alternative attempt at solving the quantum measurement problem, by modifying the dynamical laws of quantum mechanics, has been made by Ghirardi, Rimini and Weber[67] and Pearle.[156] Related work by Diósi[50] and Belavkin[16] is noteworthy. In this case, one assumes the existence of an underlying stochastic process that makes the dynamics of a quantum mechanical n-particle system nonunitary, and provokes a *spontaneous localization* process. The localization takes place at random times, with a very small rate, so that quantum systems composed of very few particles evolve according to the ordinary Schrödinger equation. On the other hand, when the system is macroscopic, and made up of a huge number of elementary constituents, the probability that the system becomes localized is practically unity.

The model is noteworthy, because an attempt is made at a unified description of micro and macroscopic systems. It should be stressed, however, that no physical explanation exists, at present, that can justify the presence of the underlying stochastic process governing the dynamics. The generalized Liouville equation they propose involves indeed localization operators and rates that are introduced *ad hoc* into the theory. Moreover, the model itself is not easily reconciled with the causality requirements of special relativity.

In the light of these considerations, it would be highly desirable to justify the postulates of the theory on a sound physical basis. There have been several attempts along this line of thought, in the past, but none of them was really satisfactory, for one reason or another. Ghirardi, Rimini, Weber and Pearle's scheme is undoubtedly an appealing model of this sort, but is still rather far from being a complete theory. Notice that it is possible to derive a nonlinear Schrödinger equation by eliminating all the variables describing the surrounding material; on the other hand, in the above-mentioned approach (as well as in earlier attempts), the stochastic process is given a fundamental status and is not supposed to be obtainable by tracing over the environmental variables.

It would be appealing to have a theory incorporating irreversible and nonunitary evolutions, as well as a unified description of the micro and macroscopic world. We wonder whether this would be close to an ultimate theory of the physical world. We feel that undertaking such a program is somewhat premature, at present. The MHS theory is rather conservative in comparison with this attempt.

4 The Einstein-Podolsky-Rosen paradox

Einstein, Podolsky and Rosen (EPR)[53] proposed in 1935 an interesting paradox in order to criticize quantum mechanics. Their argument was based on two criteria, that endeavoured to sharpen our intuitive conception of "physical reality:" (i) *every element of the physical reality must have a counter-part in the physical theory*, and (ii) *if, without in any way disturbing a system, we can predict with certainty (i.e., with probability equal to unity) the value of a physical quantity, then there exits an element of physical reality corresponding to this physical quantity.*

We will present Bohm's compact exposition[25] of the EPR paradox: Consider a dynamical system composed of two identical spin-1/2 particles in the singlet state

$$|\chi\rangle = \frac{1}{\sqrt{2}} \left[|+z\rangle_1 |-z\rangle_2 - |-z\rangle_1 |+z\rangle_2 \right], \qquad (3.35)$$

$$\sigma_z^{(i)} |\pm z\rangle_i = \pm |\pm z\rangle_i, \quad (i = 1, 2),$$

where $\sigma_z^{(i)}$ is the third Pauli matrix acting on particle #i. If, after the two particles are spatially separated, we measure the observable $\sigma_z^{(1)}$ on particle #1 and get the eigenvalue +1 (−1), then we immediately know, as a consequence of the corresponding wave function collapse, that the observable $\sigma_z^{(2)}$ of particle #2 takes the value −1 (+1). This is usually referred to as *nonlocal long-distance correlation*.

According to the EPR criterion (ii), there must be an element of physical reality corresponding to the observable $\sigma_z^{(2)}$, because we can infer its value (−1 or +1) without disturbing particle #2. This means that the z-spin component of particle #2 (i.e. the observable $\sigma_z^{(2)}$) has a definite value (−1 or +1) from the outset. This is clearly extraneous to the quantum mechanical framework.

But there is even more: Indeed, Eq. (3.35) is identically equal to

$$|\chi\rangle = \frac{1}{\sqrt{2}} \left[|+x\rangle_1 |-x\rangle_2 - |-x\rangle_1 |+x\rangle_2 \right] \qquad (3.36)$$

where $|\pm x\rangle = (|+z\rangle \pm |-z\rangle)/\sqrt{2}$ are the two eigenstates of σ_x belonging to the eigenvalues 1 and −1, respectively. Therefore, the same considerations applied before force us to conclude that $\sigma_x^{(2)}$ has a definite value (−1 or +1) from the outset. Summarizing, we are led to the contradictory conclusion that the two non-commutable observables $\sigma_z^{(2)}$ and $\sigma_x^{(2)}$ have definite values from the outset. Einstein, Podolsky and Rosen claimed that the quantum mechanical description of the physical reality given by the wave function is not complete.

Replying to this criticism, Bohr[28] stressed that we should not consider a dynamical system with a definite correlation as composed of two particles separated and independent of each other, but should rather regard a measurement on one particle as a measurement on the whole system. This property of quantum systems is often called *nonseparability* or *indivisibility*. Einstein strictly refused to accept the existence of nonlocal long-distance correlations and of nonseparability. This is the essence of the famous Einstein-Bohr debate.

This debate provoked various attempts to reformulate quantum mechanics within a classical-like framework, such as Bohm's *hidden variable* theory. See Refs. 26, 194, 49, 79, 187. In particular, there were several attempts at deriving *local* hidden variable theories, free from nonlocal long-distance correlations. As is well known, however, Bell[17] derived an inequality that experimentally discriminates quantum mechanics from all local hidden variable theories. Let us sketch Bell's theorem. Define the two dichotomic observables

$$A(a, \lambda) = \pm 1, \qquad B(b, \lambda) = \pm 1, \qquad (3.37)$$

which represent the outcome of a spin measurement on the first and second electron along directions a and b, respectively. The quantity λ represents a (set of) hidden variable(s), distributed over the space Λ with probability measure $\mu(\lambda)$, so that we can write the correlation function

$$E(a, b) = \int_{\Lambda} d\mu(\lambda) A(a, \lambda) B(b, \lambda). \qquad (3.38)$$

Notice that Λ and μ depend neither on a, nor on b, and $A(a, \lambda), B(b, \lambda)$ do not depend on b, a, respectively. This is Bell's formalization of Einstein's locality requirement.

By starting from (3.37) and (3.38) and by requiring that the observables be perfectly anticorrelated for $a = b$ [in order to obtain the same results of the singlet state (3.35) for opposite directions of spin observation], one easily gets the *Bell inequality*

$$|E(a, b) - E(a, b')| + |E(a', b) + E(a', b')| \leq 2. \qquad (3.39)$$

On the other hand, a straightforward quantum mechanical calculation yields

$$E(a, b) = \langle \chi | \sigma^{(1)} \cdot a \otimes \sigma^{(2)} \cdot b | \chi \rangle = -a \cdot b \qquad (3.40)$$

where $|\chi\rangle$ is the singlet state and $\sigma^{(i)} (i = 1, 2)$ the set of Pauli matrices. It is well known that the Bell inequality (3.39) is violated for suitable choices of the parameters a, a', b, b'.

The experiment proposed by Bell was performed by several experimental groups and the quantum mechanical predictions found correct. See for example Refs. 38, 62, 9, 159, 149, 3, 103.

There have been several attempts at understanding the EPR phenomenon from a "realistic" standpoint.[39] Unfortunately, all tentative solutions depend on what the authors define as "realism," and there is no unanimous consensus on the meaning of "realism" in quantum mechanics. Moreover, none of the solutions proposed so far can be considered satisfactory, either because they rely upon strong (and often unnatural) assumptions, or because they can be shown to be in disagreement with some of the many experiments performed so far.

One can also adopt an alternative viewpoint, and discuss the meaning of the concept of *information*. (For an interesting account of this viewpoint, see Ref. 44.) In fact, from an operational standpoint, gaining or neglecting information about the wave function can be shown to yield a change of the entropy associated with the quantum system, and one can argue that this corresponds to a genuine physical process. Although this point of view is very interesting, it is difficult, for us, to accept lightheartedly its profound implications: Indeed, at least in the EPR context, the gain or loss of information is a process that takes place only in the mind of the observer, and this should not bear any physical influence on the wave function itself. If we want to consider the collapse of the wave function as a *real* process, implying a substantial modification of the density matrix of the total system, we cannot accept the idea that the mind of the observer, by gaining or neglecting information about the quantum system, may influence its future behaviour.

In this book we shall not discuss in detail the EPR problem: Exhaustive analyses can be found in Refs. 39, 178, 47.

Chapter 4

THE MANY-HILBERT-SPACE THEORY

We have already described the basic ideas of the many-Hilbert-space theory in Section 2.4 of Chapter 1. According to these ideas, each state of the detector D (made up of many local systems) is associated with the ℓth incoming Q particle met in an experimental run. Remember that the ℓth state of the D system is in general different from other states, characterized by $\ell' \neq \ell$, due to the internal motion of the detector and other uncontrollable factors. This is true even for a single experimental run carried out under the same (macroscopic) conditions.

1 Improving the von Neumann-Wigner theory from the MHS-point of view

1.1 New detector model and WFC

Let us start by improving the description (3.7) for detector modelling. We suppose that the detector system is designed so as to yield

$$\hat{S}\left(|\chi_i\rangle \otimes |\Phi_I^{D,(\ell)}\rangle\right) = S_i^{(\ell)}|\chi_i\rangle \otimes |\Phi_i^D\rangle, \tag{4.1}$$

$S_i^{(\ell)}$ being a c-number.[a] As was mentioned before, one of the central problems at the core of the measurement theory is to evaluate the S-matrix in (3.6) for the collision process between Q and D. Here we have assumed that the only dependence on different local systems is given by the c-number $S_i^{(\ell)}$, without any effects on other factors. Note that for $S_i^{(\ell)} = 1$, this detector model goes back to the simpler one (3.7). The subscript i stands for the different channels appearing in the spectral decomposition process (1.2). Roughly speaking, we

---

[a]Note that the detection step takes place after the spectral decomposition, so that the reader can understand the reason why we replace Φ_0^D in (3.7) with $\Phi_I^{D,(\ell)}$ in (4.1). The essential difference between (3.7) and (4.1) is in the presence of $S_i^{(\ell)}$ in the latter.

have assumed by (4.1) that the detector system is not seriously altered through-out the measurement process. This way of modelling a detector is acceptable because a measurement does not destroy the detector itself, whose essential structure and function are kept unaltered. Equation (4.1) is to be understood as a sort of "uniformity" assumption for the final state of the detector. We shall generalize this detector model in Section 2.

Recalling the fundamental postulate of quantum mechanics, which gives the statistical law for the accumulated distribution over many experimental results performed under the same conditions, we can express the whole measurement process in terms of the total density matrix: For fixed ℓ, the transition is $\hat{\rho}_I^{(\ell)} \to \hat{\rho}_F^{(\ell)}$ and by averaging over ℓ we get

$$\hat{\Xi}_I \equiv \overline{\hat{\rho}_I} \longrightarrow \hat{\Xi}_F \equiv \overline{\hat{\rho}_F} = \sum_{i,j} \Delta_{ij} |\chi_i\rangle\langle\chi_j| \otimes |\Phi_i^D\rangle\langle\Phi_j^D|, \qquad (4.2)$$

where we have denoted the average over ℓ by a bar

$$\overline{(\cdots)} \equiv \frac{1}{N_p} \sum_{\ell=1}^{N_p} (\cdots)^{(\ell)}, \qquad (4.3)$$

and have defined

$$\Delta_{ij} \equiv \overline{S_i S_j^*}. \qquad (4.4)$$

In this approximation, all the effects of the detector system are simply contained in Δ_{ij}.

Let us divide Δ_{ij} into its "diagonal" and "off-diagonal" parts

$$\Delta_{ij} = d_i \delta_{ij} + \eta_{ij}(1 - \delta_{ij}), \qquad (4.5)$$

where, obviously, a nonvanishing value of the off-diagonal part represents the degree of remaining coherence. Consequently, perfect dephasing is characterized by

$$d_i = 1, \quad \eta_{ij} = 0, \qquad (4.6)$$

if we neglect leakage and absorption. In this case (4.2) becomes ($\hat{\xi}_k^Q = |\chi_k\rangle\langle\chi_k|$)

$$\hat{\Xi}_I \longrightarrow \hat{\Xi}_F = \sum_k |c_k|^2 \hat{\xi}_k^Q \otimes |\Phi_k^D\rangle\langle\Phi_k^D|, \qquad (4.7)$$

which is nothing but a WFC process of the type (3.5) with $\hat{\sigma}_{F(k)} = |\Phi_k^D\rangle\langle\Phi_k^D|$. In other words, the WFC takes place upon measurement, when the factor $S_i^{(\ell)}$, given by the interaction of the ℓth particle with the detector, becomes a random sequence with respect to ℓ. The reader can understand this situation by referring to Fig. 4.1. On the contrary, the WFC by measurement does not

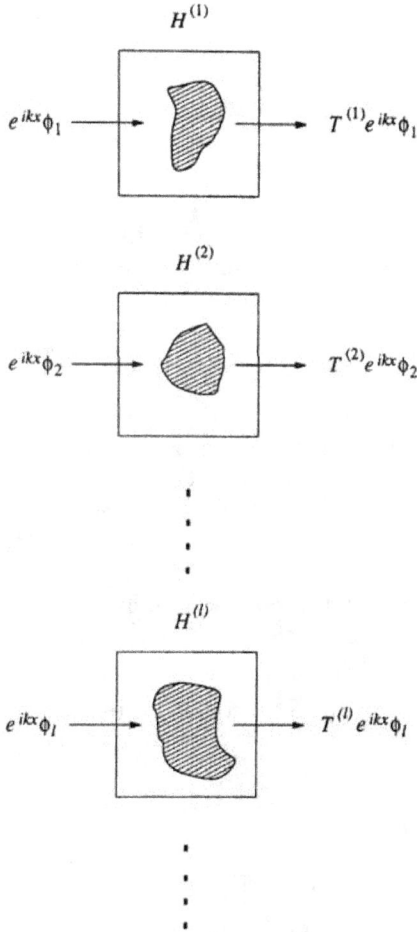

Fig. 4.1. Transmission coefficient depending on ℓ.

take place if the sequence $S_i^{(\ell)}$ has no remarkable fluctuations as a function of ℓ. For the extreme case in which $S_i^{(\ell)}$ is independent of ℓ, we have no loss of quantum coherence (and *a fortiori* no WFC) and the many Hilbert spaces, describing the detector system, reduce to a single Hilbert space, so that Wigner's "No-Go" theorem, explained in Section *2.2* of Chapter 1, holds true again. In this context, the MHS theory does not contradict Wigner's theorem, but enables one to develop the measurement theory beyond any similar "No-Go" theorem.

As was repeatedly emphasized, the average over ℓ engenders the use of a direct sum of many Hilbert spaces. Someone might think that this procedure has introduced an additional new rule into quantum mechanics. But remember that this is just a natural consequence of our careful attempt at following the actual experimental procedure, and does not imply any alteration of quantum mechanics. The general theory of quantum measurements will be formulated later (Section 2) along the same line of thought.

1.2   Yes-no experiments and decoherence parameter

We apply the theory sketched above to a yes-no experiment, like a double-slit interference or a spin measurement of the Stern-Gerlach type. See Figs. 1.3 or 1.4. In this case we have only two channels A and B (i.e. $i =$ a, b) and we place only detector D_a in channel A, as explained in Section *1.3* of Chapter 1. For simplicity, we assume that the collision between Q and D_a can be described by a one-dimensional scattering process and each incoming Q particle can be well approximated by a wave packet very close to a plane wave of wave number k. In this case, we can also assume that the collision processes in A and B are well represented by $S_a^{(\ell)} = T^{(\ell)}$ (transmission coefficient) and $S_b^{(\ell)} = 1$, without any other change. $S_{a,b}^{(\ell)}$ are just the S-matrix elements in (4.1). In this simple case, (4.1) yields

$$\hat{S}u_a \otimes \Phi_I^{D,(\ell)} = T^{(\ell)}u_a \otimes \Phi_F^D, \quad \hat{S}u_b \otimes \Phi_I^{D,(\ell)} = u_b \otimes \Phi_I^D, \qquad (4.8)$$

and

$$\Delta_{aa} = t \equiv \overline{|T|^2}, \qquad \Delta_{bb} = 1, \qquad (4.9)$$

$$\Delta_{ab} = \Delta_{ba}^* = e^{i(\arg \overline{T})}\sqrt{t(1-\epsilon)}, \qquad (4.10)$$

$$\epsilon \equiv 1 - \frac{|\overline{T}|^{2\cdot}}{\overline{|T|^2}}. \qquad (4.11)$$

These are the all elements of Δ_{ij}. We expect a transmission probability $t \simeq 1$, because a detector lets all particles smoothly go through without any considerable reflection or absorption. Obviously, this does not include the case of a "destructive" detector, that will be discussed separately later.

Look at Fig. 4.1, in which D_a will yield a transmission coefficient $S_a^{(\ell)} = T^{(\ell)}$ for the ℓth Q particle in an experimental run. We shall call ϵ, defined in (4.11), the *decoherence parameter*, because it yields an estimate of the degree of decoherence, as will be shown explicitly below [see (4.14)]: For example, we have "perfect dephasing" and "perfect interference" for $\epsilon = 1$ and $\epsilon = 0$, respectively. In particular, the intermediate cases $0 < \epsilon < 1$ are very important, because we can quantitatively define an *imperfect measurement* via the value of the decoherence parameter. In this case, there is still a residual coherence, but the total state partially tends to become a mixed state, so that, so to speak, the total system is in a partially decohered state, i.e. an incompletely mixed state. We would be unable to deal with such an imperfect measurement by means of a measurement theory based on the naive WFC, specified by a sudden change of the projection type (1.7). By contrast, we can tackle this situation within the theoretical framework of the MHS theory. This is one of the most remarkable merits of the MHS theory. The decoherence parameter is a sort of *order parameter* in a sense to be explained later. Here we should notice that a quantum "imperfect measurement" is different from the "incomplete measurement" described as follows: Consider, for example, the momentum measurement in Fig. 1.2, realized by making use of perfect detectors D_is. If we remove some detectors or cover them with opaque glasses, the measurement is not complete, but in a different sense from the above notion of "imperfect measurement." The two cases should not be confused with each other.

In the above yes-no experiment, the whole process is specified by the following final state

$$\hat{\Xi}_{\mathrm{F}t}^{\mathrm{tot}} = \frac{1}{2} \left[\overline{|T|^2} \hat{\xi}_{\mathrm{a}}^{\mathrm{Q}} \otimes |\Phi_{\mathrm{F}t}^{\mathrm{D}}\rangle\langle\Phi_{\mathrm{F}t}^{\mathrm{D}}| + \hat{\xi}_{\mathrm{b}}^{\mathrm{Q}} \otimes |\Phi_{\mathrm{I}t}^{\mathrm{D}}\rangle\langle\Phi_{\mathrm{I}t}^{\mathrm{D}}| \right.$$
$$\left. + \overline{T} e^{-i(E_{\mathrm{a}} - E_{\mathrm{b}})t/\hbar} |\chi_{\mathrm{a}}\rangle\langle\chi_{\mathrm{b}}| \otimes |\Phi_{\mathrm{F}t}^{\mathrm{D}}\rangle\langle\Phi_{\mathrm{I}t}^{\mathrm{D}}| + \mathrm{h.c.} \right], \qquad (4.12)$$

where h.c. stands for Hermitian conjugate. Notice that the first two terms of the r.h.s., in the case of a perfect measurement ($\epsilon = 1$, i.e. $\overline{T} = 0$), represent the WFC in a negative-result measurement. Therefore the MHS theory breaks through the barrier of Wigner's objection and explicitly realizes the WFC in the negative-result-measurement case. Originally, this was one of the main motivations that led to the formulation of the MHS theory.

We remark that $\Phi_{\mathrm{F}}^{\mathrm{D}}$ in the r.h.s. of the first formula in (4.8) may be replaced by $\Phi_{\mathrm{I}}^{\mathrm{D}}$ for a more simplified model of detector, yielding only a c-number $T^{(\ell)}$ without any change of state. For such a detector we have

$$\hat{\Xi}_{\mathrm{F}t}^{\mathrm{tot}} = \frac{1}{2} \left[\overline{|T|^2} \hat{\xi}_{\mathrm{a}}^{\mathrm{Q}} + \hat{\xi}_{\mathrm{b}}^{\mathrm{Q}} + \overline{T} e^{-i(E_{\mathrm{a}} - E_{\mathrm{b}})t/\hbar} |\chi_{\mathrm{a}}\rangle\langle\chi_{\mathrm{b}}| + \mathrm{h.c.} \right] \otimes |\Phi_{\mathrm{I}t}^{\mathrm{D}}\rangle\langle\Phi_{\mathrm{I}t}^{\mathrm{D}}|. \quad (4.13)$$

Remember that even for (4.8), with $\Phi_F^D \neq \Phi_I^D$, we obtain the same final state if we observe the Q particles in the final channel 0 outside the D_a region and assume that detector D_a goes back to the initial state after some recovery time.

In order to measure the degree of completeness of the measurement, which is reflected in the value of the decoherence parameter, we simply have to observe the interference by another (perfect) detector D_0 located in the final channel, as shown on the right of Fig. 1.4(b). The final state is now described by (4.13) in the final channel (outside the D_a region). By assuming an unlimited precision on the point-like detection by D_0, we easily obtain the final probability of detecting Q particles by D_0

$$P = \frac{1}{2}\left[1 + t + 2\sqrt{t(1-\epsilon)}\cos\left(\arg\overline{T}\right)\right],\tag{4.14}$$

in the case of $|\langle\chi_b|\chi_a\rangle| = 1$, where, for simplicity, all additional phases have been included in the argument of $\overline{T}$. The above formula enables us to infer the value of the decoherence parameter ϵ by observing interference at D_0. Thus we can judge whether the instrument D_a works well or not as a detector, by estimating the value of ϵ. Notice that we are not directly concerned as to whether D_a actually yields a signal or not as a result of the interaction. In this sense, we are looking at its property as a *dephaser*. Keep this remark in mind throughout the following discussion.

It is instructive and interesting to compute the *visibility* of the interference pattern

$$V = \frac{P_{\text{MAX}} - P_{\min}}{P_{\text{MAX}} + P_{\min}}.\tag{4.15}$$

From (4.14), one easily obtains

$$V_{\text{MHS}} = V_0\sqrt{1-\epsilon} = 2\frac{\sqrt{t(1-\epsilon)}}{1+t},\tag{4.16}$$

where $V_0 = 2\sqrt{t}/(1 + t)$ is the standard visibility, obtained in absence of fluctuations. Once again we see that the coherence between the two branch waves is totally lost for $\epsilon = 1$, in which case the visibility is zero.

Notice that the notion of quantum mechanical coherence requires an *operational* definition. In order that the average (4.3) be meaningful, one must accumulate many ($N_p \gg 1$) events in an experimental run. If N_p is not very large, one is not able to give an *operationally meaningful* definition of (loss of) coherence.

This is best clarified by considering the limiting situation of a perfect measurement, where

$$|\overline{T}|^2 = \frac{1}{N_p^2}\sum_{\ell,\ell'}T^{(\ell)}T^{(\ell')*} = \frac{1}{N_p}\left(t + \frac{1}{N_p}\sum_{\ell\neq\ell'}T^{(\ell)}T^{(\ell')*}\right) \simeq \frac{1}{N_p}t,\tag{4.17}$$

because the summation $\sum_{\ell \neq \ell'}$ is very small, in the case of a perfect measurement (complete decoherence). Thus we obtain, for a perfect measurement,

$$\epsilon = 1 - \frac{|\overline{T}|^2}{\overline{|T|^2}} \simeq 1 - \frac{1}{N_p}. \qquad (4.18)$$

Therefore, there appear to be two distinct criteria for a complete loss of quantum mechanical coherence. First, one needs a "good" detector, characterized by a huge number of degrees of freedom, that is, a huge number of local systems (this is neither a necessary, nor a sufficient condition, as we shall see in Chapters 7 and 9). Second, Eq. (4.18) requires that one performs the experiment many times ($N_p \gg 1$), in order to be able to ascertain whether the quantum mechanical coherence is lost.

It goes without saying that the above conclusion, drawn in the case of a perfect measurement (complete loss of coherence), holds true in every case ($0 \leq \epsilon \leq 1$), and even when $\epsilon \to 0$. If N_p is not much larger than 1 (i.e. there are not so many events in one experimental run) in general one may not be able to give an operationally meaningful definition of (loss of) coherence.

The formula (4.14) for the probability of finding Q particles at D_0 corresponds to the observation of Q particles without any specification on position. In the case here considered (yes-no experiments with two branch waves running in channels A and B and recombined in the final channel 0), we can formulate the probability of finding a sharp value ξ of an arbitrary Q-observable $\hat{F}$ as

$$P_\xi = (\psi_a, \hat{I}(\xi)\psi_a) + (\psi_b, \hat{I}(\xi)\psi_b) + 2\text{Re}(\psi_b, \hat{I}(\xi)\psi_a), \qquad (4.19)$$

where

$$\hat{I}(\xi) \equiv \delta(\hat{F} - \xi). \qquad (4.20)$$

Note that we have used the same notation ψ_a to express both the original wave function and the modified wave function, after the passage through D_a. One may find it convenient and more realistic to consider, instead of the above δ-function, the projection operator for the observable $\hat{F}$

$$\hat{\bar{I}}(\xi) \equiv \int_{\mathcal{D}(\xi)} d\xi' \delta(\hat{F} - \xi'), \qquad \hat{\bar{I}}^2(\xi) = \hat{\bar{I}}(\xi), \qquad (4.21)$$

where $\mathcal{D}(\xi)$ stands for a domain centered around ξ with width Δ. This kind of treatment (i.e. the integration over a finite domain) will always be necessary for those observables with continuous eigenvalues, however, we prefer to use (4.20) because of its simplicity and the sharp specification of ξ. (See also the discussion at the end of Section 2.2 of this chapter.) Upon integration over ξ (notice that $\int \hat{I}(\xi)d\xi = 1$), (4.19) becomes

$$\frac{1}{2}\left[1 + t + 2|(\psi_b, \psi_a)|\sqrt{t(1-\epsilon)}\cos(\arg \overline{T})\right], \qquad (4.22)$$

when both wave packets ψ_a and ψ_b are very close to plane waves. In the case in which the two wave packets do not overlap [i.e. $(\psi_b, \psi_a) \sim 0$], the interference term disappears as was clearly shown in the neutron experiments performed by the Rauch-Werner group.[40,93,87,162] They claimed that such a situation does not necessarily imply a loss of coherence and experimentally showed that interference can be beautifully recovered when the components with wave numbers nearly equal to k are selected and superposed. This point is repeatedly emphasized in this book and will be analyzed again in Section *1.3* of Chapter 6. The experimental setup[40] corresponds to choosing momentum as $\hat{F}$ and $\hat{I} = \delta(\hat{F} - k)$. For this choice the factor $(\psi_b, \hat{I}\psi_a)$ never vanishes, even for non-overlapping [i.e. $(\psi_b, \psi_a) \sim 0$] wave packets, and yields a clear interference pattern.

It is conceivable that we will even be able to draw a kind of time series for the passage of one incident wave packet from future interferometric experiments. (One should not forget that there always exists an ensemble of incident particles corresponding to the single wave packet.) If we can observe such a time series and study its details by means of Fourier analysis, we will be able to display a remarkable difference in the Fourier spectrum for the cases with and without signals. This would be a new possible way of detecting signals. Equation (4.19) is a general formula describing interference phenomena and is to be regarded as a starting point to analyze new detection techniques.

1.3   Mesoscopic phenomena

We have introduced the notion of *imperfect measurement*, characterized by a decoherence parameter $0 < \epsilon < 1$, representing a partially decohered state of the whole system. Such a physical situation is very similar to those observed in the so-called *mesoscopic* phenomena, which have come to our attention during the last decade.[199,182,126,188] Actually, mesoscopic physics is now rapidly developing and becoming one of the most important topics in condensed matter physics. Both mesoscopic phenomena and imperfect measurements have a common physical basis. For this reason, the theory of measurement must be able to quantitatively analyze imperfect measurements in order to deal with mesoscopic phenomena.

Recently, Monroe *et al.*[126] claimed to have realized a Schrödinger's cat state as a mesoscopic phenomenon. Originally, such a state is given by an entangled superposition of the type (3.10). Suppose that we have an appropriate macroscopic system D characterized by two "main" states, say $|\phi_{a,b}\rangle$, in a one-to-one correspondence with two internal states, say $|\pm\rangle$, of a microscopic system Q. Identify Q and D with the radioactive material and the cat in the Schrödinger cat paradox, respectively, so that the state of the total system

(Q+D) reads

$$|\Psi^{Q+D}\rangle = \frac{1}{\sqrt{2}} \left[|\phi_a\rangle \otimes |+\rangle + |\phi_b\rangle \otimes |-\rangle \right].$$ (4.23)

The above-mentioned group experimentally realized $|\phi_{a,b}\rangle$ as two localized states of D, keeping at the same time a certain one-to-one correspondence with two internal states $|\pm\rangle$. Therefore, if one finds D in one of the two localized states, one can safely conclude that the microscopic system is in the corresponding internal state. However, obviously, the r.h.s. is nothing but the final state of the von Neumann process (3.8), and is a pure state. Of course, we cannot exactly realize such a perfect pure state involving a macroscopic system. The experimental group claimed that this kind of phase-correlated state was realized as a mesoscopic phenomenon, but not as a perfectly pure state. From the MHS-theoretical point of view, we can say that the above "cat" state is realized in the same sense as an *imperfect* measurement. In other words, this experiment brought the system to a sort of mesoscopic state, i.e. in a partially decohered state between the two components (corresponding to the "alive" and "dead" states of the cat). In this sense, Monroe *et al.*'s claim to have actually realized a Schrödinger's cat state is considered to be partially justified. Nevertheless, their experiment is very interesting and promising: In the future we should be able to perform quantum measurements via this kind of mesoscopic state.

2 General theory of measurements

The above theoretical procedure was based on the simple detector model (4.1) and on the computation of the bar-average (4.3). In this section, we shall dispense with the simplifying assumption (4.1), by making use of the general definition of S-matrix in the theory of scattering, and shall replace the bar-average (4.3) with the statistical ensemble average. We shall endeavour to formulate the general theory of measurement by means of such a general theoretical procedure.

2.1 Macroscopic instruments

In the preceding section, we have introduced the bar-average over many different local systems, met by different Q particles of the same incident beam during an experimental run. Each local system is usually specified by quantities such as number of constituents, volume and so on, all of which are a sum of physical quantities pertaining to individual constituents and are to be represented by macroscopic state variables after some averaging procedure. The bar-average is to be taken over such quantities. We recall the following results of the conventional theory of multiple scattering: In general the S-matrix, and therefore the quantity $S_i^{(\ell)}$ in the above description, has a phase factor $e^{i\Theta^{(\ell)}}$, with

$$\Theta^{(\ell)} = -\sum_{n=1}^{N^{(\ell)}} \kappa_n^{(\ell)} a_n^{(\ell)}, \qquad (4.24)$$

where $\kappa_n^{(\ell)}$'s are constants (generally complex numbers) depending on the collision dynamics, $N^{(\ell)}$'s numbers of multiple scatterings and $a_n^{(\ell)}$'s spacings among different constituents. We must take the average of $S_i^{(\ell)} S_j^{(\ell)*}$ (depending on $N^{(\ell)}$, $\kappa_n^{(\ell)}$ and $a_n^{(\ell)}$) with respect to ℓ, in order to obtain Δ_{ij} [see (4.4)].

Since different Q particles meet different local systems in different microstates, we are allowed to replace the bar-average with the *statistical ensemble average* over all the possible microstates of the detector

$$\overline{\cdots} = \langle \cdots \rangle, \qquad \text{for } N_p \gg 1, \qquad (4.25)$$

where we denoted the statistical ensemble average with $\langle \cdots \rangle$. Equation (4.25) is to be regarded as a sort of *ergodic assumption*, even though it is not exactly equivalent to the classical ergodic theorem.[b] Notice that the ergodic assumption is justified only when N_p is very large.

We have first introduced N_p as the total number of incoming Q particles brought by the incident beam in an experimental run. The bar-average in the l.h.s. of (4.25) is taken over many local systems of the detector, which interact with the Q particles belonging to the incident beam. On the other hand, we know that a single Q particle wave packet, in an individual event, will meet many local systems during its passage inside the detector, because the macroscopic detector is made up of a huge number of local systems. Taking this fact into account, we are naturally led to another (somewhat different) kind of ergodic assumption: The average over the local systems can be equated to the above-mentioned bar-average. Note that in this case N_p represents the total number of relevant local systems met by one Q particle in the detector. Thus we can use the relation (4.25) for the passage of a single Q particle wave packet inside the macroscopic detector, and eventually obtain the WFC (3.5) even for an individual event. On this basis, and in this ergodic sense, we can talk about a *statistical* and an *individual* WFC by means of the same formula (3.5). At any rate, one should not forget that we need to perform the accumulation of many experimental results in order to experimentally observe (3.5) itself.

In order to replace the bar-average of the phase Θ, given by (4.24), with its statistical ensemble average, we divide a_n into its average a_0 and a fluctuating

---

[b] In classical physics, the usual formulation of the ergodic theorem states that the long-time average of an observable can be replaced with its ensemble average over the phase space. (Incidentally, in general, this is not easy to prove mathematically.) On the contrary, we do not exactly know what the quantum version of the theorem is.

component α_n as follows

$$a_n = a_0 + \alpha_n, \quad \langle \alpha_n \rangle = 0, \quad \langle \alpha_n \alpha_{n'} \rangle = (\Delta a)^2 \delta_{nn'}, \tag{4.26}$$

where $\langle \cdots \rangle$ stands for the statistical ensemble average over the random variable α_n. For simplicity, we assume that κ_ns have no fluctuation (it would be very easy to take also these fluctuations into account). Thus, for fixed N, we can rewrite the phase as

$$\Theta_N = -N\kappa_0(a_0 + \zeta), \tag{4.27}$$

where

$$\kappa_0 \equiv \frac{1}{N} \sum_{n=1}^{N} \kappa_n, \qquad \zeta \equiv \frac{1}{N\kappa_0} \sum_{n=1}^{N} \kappa_n \alpha_n. \tag{4.28}$$

Remember that N is the total number of constituents (or of local systems), and ζ is a random variable distributed around zero, belonging to a certain statistical ensemble. For very large N, the central limit theorem yields a Gaussian law for the probability distribution of ζ:[c]

$$W(\zeta) = \frac{1}{\sqrt{2\pi \langle \zeta^2 \rangle}} \exp\left(-\frac{\zeta^2}{2\langle \zeta^2 \rangle}\right), \tag{4.29}$$

where

$$\langle \zeta^2 \rangle = \frac{(\Delta a)^2 \overline{\kappa^2}}{N\kappa_0^2}, \qquad \overline{\kappa^2} \equiv \frac{1}{N} \sum_{n=1}^{N} \kappa_n^2. \tag{4.30}$$

Notice that we required κ_0 and $\overline{\kappa^2}$ to have definite values for very large N. In the above formulae, Δa and a_0 are considered to be of the order of the atomic or molecular size. Observe also that ζ is distributed in a limited region.

Now we can explicitly write the statistical ensemble average of any quantity $G_N(\zeta, \cdots)$ as follows:

$$\langle G_N \rangle = \int d\zeta\, W(\zeta) \langle\langle G_N(\zeta, \cdots) \rangle\rangle \equiv W \cdot G_N, \tag{4.31}$$

where $\langle\langle G_N(\zeta, \cdots) \rangle\rangle$ stands for the statistical average over the microstates of the local system for fixed N and ζ.

If we choose $G_N(\zeta, \cdots) = e^{i\Theta_N}$, the above statistical ensemble average becomes

$$\langle e^{i\Theta_N} \rangle = e^{-iN\kappa_0 a_0} W \cdot e^{-iN\kappa_0 \zeta} = e^{-iN\kappa_0 a_0} e^{-N\overline{\kappa^2}(\Delta a)^2/2} \xrightarrow{N \to \infty} 0, \tag{4.32}$$

---

[c]Strictly speaking, the quantity which follows the Gaussian law in the infinite-N limit is not ζ, but $\zeta' \equiv \sqrt{N}\zeta$. The following formulae are expressed, for notational simplicity, in terms of ζ and are, of course, valid for very large N.

because

$$W \cdot e^{-iN\kappa_0\zeta} = \int_{-\infty}^{\infty} e^{-iN\kappa_0\zeta} W(\zeta) d\zeta = e^{-N\overline{\kappa^2}(\Delta a)^2/2}, \qquad (4.33)$$

where we assumed κ_0 and $\overline{\kappa^2}$ to be real, for simplicity. By generalizing (4.32), we can easily show that, for any bounded $\langle\langle G_N \rangle\rangle$ with a finite support as a function of ζ,

$$\lim_{N\to\infty} W \cdot e^{-iN\kappa_0\zeta} G_N = \lim_{N\to\infty} \int_{-\infty}^{\infty} d\zeta W(\zeta) e^{-iN\kappa_0\zeta} \langle\langle G_N \rangle\rangle = 0, \qquad (4.34)$$

by virtue of the Riemann-Lebesgue lemma. We shall see later that this mathematical theorem is a very powerful tool in order to derive the WFC when

$$N\overline{\kappa^2}(\Delta a)^2 \gg 1. \qquad (4.35)$$

For finite N, the representation space belongs to the discrete-superselection-rule space (see Section 4 of Chapter 2), but in the infinite N limit, the space tends to the continuous-superselection-rule space as indicated in (2.85). We shall describe in the following this transition of the representation space.

Before describing the details of such a procedure, let us consider the possibility of fluctuations of N around N_0 with width ΔN. In this case, we have to introduce an averaging procedure over N, in addition to the above average (4.31) over ζ (we use the same symbol W for simplicity)

$$W \cdot G \equiv \sum_{N \in I(N_0, \Delta N)} W_N \langle G_N \rangle = \sum_{N \in I(N_0, \Delta N)} W_N \int d\zeta W(\zeta) \langle\langle G_N(\zeta) \rangle\rangle, \qquad (4.36)$$

where $I(N_0, \Delta N)$ is a set of possible values of N, with average N_0 and standard deviation ΔN, and W_N is the statistical weight with respect to N ($\sum_N W_N = 1$).

Define a discrete length variable and its average by

$$l_N = \frac{1}{\kappa_0} \sum_{n=1}^{N} \kappa_n a_n, \quad \langle l_N \rangle = N_0 a_0, \qquad (4.37)$$

respectively, so that its mean square deviation and relative ratio are given by

$$\langle (\Delta l_N)^2 \rangle = (\Delta N)^2 a_0^2 + N_0 (\Delta a)^2 \frac{\overline{\kappa^2}}{\kappa_0^2}, \qquad (4.38)$$

$$\frac{\langle (\Delta l_N)^2 \rangle}{\langle l_N \rangle^2} = \frac{(\Delta N)^2}{N_0^2} + \frac{\overline{\kappa^2}}{N_0 \kappa_0^2} \frac{(\Delta a)^2}{a_0^2}. \qquad (4.39)$$

Thus we can replace the discrete quantity l_N with a continuous parameter l in the infinite N limit, if we perform a micro-macro scale transformation and regard the ratio (4.39) as an infinitesimal quantity. In correspondence to the scale transformation, the above-mentioned averaging procedure is expressed in the following integral form

$$\mathcal{W} \cdot G = \int dl W(l) \langle\langle G \rangle\rangle(l), \tag{4.40}$$

where $W(l)$ and $\langle\langle G \rangle\rangle(l)$ are the scale-transformed quantities $W_N W(\zeta)$ and $\langle\langle G \rangle\rangle_N$, respectively. Since N is very large, we can safely use the Gaussian distribution

$$W(l) = \frac{1}{\sqrt{2\pi(\Delta l)^2}} \exp\left(-\frac{(l-l_0)^2}{2(\Delta l)^2}\right), \tag{4.41}$$

where $l_0 = \langle l \rangle$ is the limit of $\langle l_N \rangle = N_0 a_0$ for $N_0 \to \infty$ and $a_0 \to 0$, keeping their product finite, and $\Delta l = \lim \Delta l_N$, under the same conditions. Obviously, we are now using a continuous-superselection-rule space, $\int \oplus d\mu(\zeta)\mathcal{H}(\zeta)$, for the mathematical representation of the macroscopic instrument, as suggested in Section 4 of Chapter 2. (One can choose not only the above-mentioned length variable, as a continuous macroscopic parameter, but also volume, electric current or magnetic flux, depending on the physical problem investigated.)

The density matrix for finite N

$$\hat{\rho}^{\mathrm{D}} = \sum_{N \epsilon I(N_0, \Delta N)} W_N \langle \hat{\rho}_N^{\mathrm{D}} \rangle \tag{4.42}$$

tends to the following expression in the infinite N limit,

$$\hat{\sigma}^{\mathrm{D}} \equiv \mathcal{W} \cdot \hat{\rho}^{\mathrm{D}} = \int dl W(l) \langle\langle \hat{\rho}^{\mathrm{D}} \rangle\rangle(l), \tag{4.43}$$

where $\langle\langle \hat{\rho}^{\mathrm{D}} \rangle\rangle(l)$ is the density matrix of a local system of D characterized by a sharp value of the macroscopic size parameter l. The quantity $\hat{\sigma}^{\mathrm{D}}$ is just the density matrix averaged over many events pertaining to the same macroscopic experimental conditions. This is the *mathematical description of a macroscopic instrument*. In this representation, we have to replace (4.33) with

$$\mathcal{W} \cdot e^{-i\kappa_0(l-l_0)} = \int_{-\infty}^{\infty} e^{-i\kappa_0(l-l_0)} W(l) dl = e^{-\frac{1}{2}\kappa_0^2(\Delta l)^2}. \tag{4.44}$$

Equation (4.44) and integrals of the type

$$\mathcal{W} \cdot e^{-i\kappa_0(l-l_0)} G = \int_{-\infty}^{\infty} e^{-i\kappa_0(l-l_0)} W(l) \langle\langle G(l) \rangle\rangle dl \simeq 0, \tag{4.45}$$

for $\langle\langle G(l) \rangle\rangle$ with a finite support, become very small by virtue of the Riemann-Lebesgue lemma, if

$$\kappa_0^2(\Delta l)^2 \gg 1. \tag{4.46}$$

Equation (4.45) under (4.46) is to be compared with (4.34) under (4.35) in the case of nonfluctuating N. We shall see later that one can obtain the WFC under this condition, while no WFC is realized if (4.46) is not satisfied. The averaging procedure just described is both subtle and interesting, from the mathematical point of view. In the context of the many-Hilbert-space approach, it has been discussed by Araki[6] and Stratonovich and Belavkin.[183]

Note that every local system is characterized by a finite size l, which is very small on a macroscopic scale but very large on a microscopic scale. The scale transformation, accompanied by the integration with weight $W(l)$, is regarded as a coarse-graining procedure, bridging the gap between the two different levels of description, the microscopic and the macroscopic one.

2.2   Yes-no experiment and destructive detector

Let us first consider the case in which a Q wave function is split into two states, corresponding to two possible routes A and B in a yes-no experiment. See, for example, Fig. 1.3 or 1.4. We shall place a detector D along the first route A. This is one of the simplest arrangements for a measuring apparatus, and will enable us to make a useful comparison with the general formulae in the following subsection.

The density matrix corresponding to the final state of the spectral decomposition $\psi_0 \rightarrow \psi_I = \psi_a + \psi_b$ can be written as

$$\hat{\rho}_I = \sum_{i,j=a,b} |\psi_i\rangle\langle\psi_j|. \tag{4.47}$$

The above notation is trivial in the case of an interference experiment of the Young type; in the case of a Stern-Gerlach experiment one sets $|\psi_{a,b}\rangle = |\chi_{a,b}\rangle = |u_{a,b}\rangle|\phi_{a,b}\rangle$ (see Section *1.3* of Chapter 1). According to the discussion of Section 1 (Section *1.2*, in particular), we have to add the superscript ℓ to all the relevant quantities in the measuring process and take the average of the total density matrix over ℓ, because the ℓth Q particle in an experimental run will meet the ℓth local system in a microscopic state described by $\hat{\rho}_I^{D(\ell)}$ and will undergo a transition described by the S-matrix, whose elements are $\hat{S}_a^{(\ell)}$ and $\hat{S}_b^{(\ell)} = 1$. According to the formal theory of scattering, we have to describe the elementary interaction between Q and D (or, in general, some local system of the latter) as follows:

$$e^{-i\hat{H}t/\hbar} \xrightarrow{t \rightarrow \infty} e^{-i\hat{H}_0 t/\hbar}\hat{S}, \tag{4.48}$$

where $\hat{H}$ and $\hat{H}_0 = \hat{H}_0^Q + \hat{H}_0^D$ are the total and free Hamiltonians of the system composed of Q and D, respectively, $\hat{H}_0^{Q(D)}$ being the free Hamiltonian of system Q (D). Hence, the accumulated distribution on the screen, during an experimental run, is expressed by the Q component of the asymptotic density matrix

$$\hat{\Xi}^{\text{tot}}_{\text{I}t} \equiv \frac{1}{N_p} \sum_{\ell=1}^{N_p} e^{-i\hat{H}^{(\ell)}t/\hbar} \hat{\rho}^{Q}_{\text{I}} \otimes \hat{\rho}^{D(\ell)}_{\text{I}} e^{i\hat{H}^{(\ell)}t/\hbar} \xrightarrow{t\to\infty} \hat{\Xi}^{\text{tot}}_{\text{F}t} = \sum_{i,j=a,b} \hat{\Xi}^{ij}_{\text{F}t}, \qquad (4.49)$$

where

$$\hat{\Xi}^{ij}_{\text{F}t} = \frac{1}{N_p} \sum_{\ell=1}^{N_p} e^{-i\hat{H}^{(\ell)}_0 t/\hbar} \hat{S}^{(\ell)} \{|\psi_i\rangle\langle\psi_j| \otimes \hat{\rho}^{D(\ell)}_{\text{I}}\} \hat{S}^{(\ell)\dagger} e^{i\hat{H}^{(\ell)}_0 t/\hbar}. \qquad (4.50)$$

One of the most important tasks of the measurement theory is to calculate the operator $\hat{S}^{(\ell)}$ in the measurement process; however, this problem can be very difficult to solve. We shall give some explicit solutions by analyzing several solvable detector models in Chapter 5. Rather than providing exact solutions, most measurement theories introduce simplifying assumptions regarding the effect of the detector. Here, however, we prefer to respect the general structure of $\hat{S}^{(\ell)}$ for the collision of a quantum mechanical particle with a target of finite size,[121] yielding

$$\hat{S}^{(\ell)} = e^{i\hat{\Theta}^{(\ell)}} \hat{S}^{(\ell)}_{\text{ch}} e^{i\hat{\Theta}^{(\ell)}} \qquad (4.51)$$

in the channel representation, where $\hat{\Theta}^{(\ell)}$ is a diagonal matrix whose matrix elements are given, for example, by (4.27), and $\hat{S}^{(\ell)}_{\text{ch}}$ describes channel couplings. With this general structure in mind, we proceed to discuss the measurement process.

As mentioned in the preceding subsection, we can replace the bar-average with the statistical average over all the possible microstates of the detector, on the basis of the ergodic hypothesis (4.25). Therefore, we can rewrite $\hat{\Xi}^{ij}_{\text{F}t}$ as

$$\hat{\Xi}^{ij}_{\text{F}t} = \mathcal{W} \cdot e^{-i\hat{H}_0 t/\hbar} \hat{S} \{|\psi_i\rangle\langle\psi_j| \otimes \hat{\rho}^{D}_{\text{I}}\} \hat{S}^\dagger e^{i\hat{H}_0 t/\hbar}, \qquad (4.52)$$

where $\mathcal{W}$ is the statistical ensemble average defined in (4.33) for the case of discrete and nonfluctuating N, in (4.36) for discrete and fluctuating N and in (4.40) for the continuous case.

Remembering that the interaction takes place only between ψ_a and D_a, we can write

$$e^{-i\hat{H}_0 t/\hbar} \hat{S} \{|\psi_a\rangle\langle\psi_a| \otimes \hat{\rho}^{D}_{\text{I}}\} \hat{S}^\dagger e^{i\hat{H}_0 t/\hbar} = |\psi_{\text{aF}t}\rangle\langle\psi_{\text{aF}t}| \otimes \hat{\rho}^{D}_{\text{F}t}, \qquad (4.53)$$

$$e^{-i\hat{H}_0 t/\hbar} \hat{S} \{|\psi_b\rangle\langle\psi_b| \otimes \hat{\rho}^{D}_{\text{I}}\} \hat{S}^\dagger e^{i\hat{H}_0 t/\hbar} = |\psi_{\text{b}t}\rangle\langle\psi_{\text{b}t}| \otimes \hat{\rho}^{D}_{\text{I}t}, \qquad (4.54)$$

$$e^{-i\hat{H}_0 t/\hbar} \hat{S} \{|\psi_a\rangle\langle\psi_b| \otimes \hat{\rho}^{D}_{\text{I}}\} \hat{S}^\dagger e^{i\hat{H}_0 t/\hbar} = |\psi_{\text{aF}t}\rangle\langle\psi_{\text{b}t}| \otimes (\hat{S}\hat{\rho}^{D}_{\text{I}t}), \qquad (4.55)$$

$$e^{-i\hat{H}_0 t/\hbar} \hat{S} \{|\psi_b\rangle\langle\psi_a| \otimes \hat{\rho}^{D}_{\text{I}}\} \hat{S}^\dagger e^{i\hat{H}_0 t/\hbar} = |\psi_{\text{b}t}\rangle\langle\psi_{\text{aF}t}| \otimes (\hat{\rho}^{D}_{\text{I}t}\hat{S}^\dagger), \qquad (4.56)$$

where $|\psi_{aFt,bt}\rangle = e^{-i\hat{H}_0^Q t/\hbar}|\psi_{aF,b}\rangle$ $(|\psi_{aF,b}\rangle = \hat{S}|\psi_{a,b}\rangle$ standing for the Q state after the measurement) and $\hat{\rho}_{Ft,It}^D = e^{-i\hat{H}_0^D t/\hbar}\hat{\rho}_{F,I}^D e^{i\hat{H}_0^D t/\hbar}$. Strictly speaking, the above relations are not completely correct, because $\hat{S}$ acts on both D and Q, provoking entanglement between them. However, we prefer to use the above (slightly incorrect) expressions in order to elucidate the effect of the measurement process.

Let us now consider the case of a destructive detector, in which the detector D_a can absorb the Q particle in the course of the measurement process. For simplicity, we shall restrict our attention to the simple case of two channels, "with" and "without" the Q particle. Taking into account the structure of $\hat{S}$, as given by (4.51), we can write

$$|\psi_{aF}\rangle = T|\psi_a\rangle + T'|0\rangle, \qquad (4.57)$$

where

$$T = e^{2i\Theta_{el}}\langle\psi_a|\hat{S}_{ch}|\psi_a\rangle, \quad T' = e^{i(\Theta_{el}+\Theta_{inel})}\langle 0|\hat{S}_{ch}|\psi_a\rangle. \qquad (4.58)$$

Here $|\psi_a\rangle$ and $|0\rangle$ (both normalized to 1) stand for 1-particle and 0-particle states, respectively. Note that T is the same transmission coefficient given in Section 1.2 (apart eventually from a trivial factor), and that T' is written in terms of $\langle 0|\hat{S}_{ch}|\psi_a\rangle$ $(|\langle 0|\hat{S}_{ch}|\psi_a\rangle|^2$ being the absorption probability). Notice also that T' involves the same phase shift as T (i.e. Θ_{el}), plus another one (Θ_{inel}). Recalling the arguments in the preceding subsection, we write the elastic part of the phase shift as

$$\Theta_{el} = -N\kappa_0(a_0 + \zeta), \qquad (4.59)$$

on the basis of the conventional theory of multiple scattering. We can also assume that T' has a structure similar to (4.59). For a detector yielding a random distribution of these phase shifts, we expect that $\overline{T} = W \cdot T = 0$ and $\overline{T'} = W \cdot T' = 0$ in the infinite N limit. As a matter of fact, the above structure of T, T', Θ_{el} and Θ_{inel} enables us to have integrals of the type (4.34) (in the case of discrete and nonfluctuating N) or (4.45) (in the continuous case), for all the off-diagonal parts of the total density matrix. For example, we have

$$\int_{-\infty}^{\infty} e^{-i\overline{\kappa}_0(l-l_0)}W(l)F(l)dl \qquad (4.60)$$

in the continuous case, where $F(l)$ is a smooth function of l with a finite support. Consequently, all the off-diagonal parts vanish for $N\overline{\kappa^2}(\Delta a)^2 \gg 1$ or $\kappa_0^2(\Delta l)^2 \gg 1$, as we have seen in (4.32) or (4.45), yielding the WFC. In this case, we can estimate the value of the *decoherence parameter* by

$$\epsilon = 1 - \exp[-N\overline{\kappa^2}(\Delta a)^2/2] \quad \text{or} \quad \epsilon = 1 - \exp[-\kappa_0^2(\Delta l)^2/2]. \qquad (4.61)$$

In the ideal case $\epsilon = 1$, we have

$$\hat{\Xi}_{Ft}^{ab} = \hat{\Xi}_{Ft}^{ba} = 0. \tag{4.62}$$

In order to display better the occurrence of WFC, let us write down explicitly the measurement process (4.49), under the condition $\epsilon = 1$:

$$\hat{\Xi}_{It}^{tot} \xrightarrow{t \to \infty} \hat{\Xi}_{Ft}^{tot} = \hat{\Xi}_{Ft}^{aa} + \hat{\Xi}_{Ft}^{bb}, \tag{4.63}$$

where

$$\hat{\Xi}_{Ft}^{aa} = \hat{\xi}_{aFt}^Q \otimes \hat{\sigma}_{Ft}^D, \quad \hat{\Xi}_{Ft}^{bb} = \hat{\xi}_{bIt}^Q \otimes \hat{\sigma}_{It}^D \tag{4.64}$$

with

$$\hat{\xi}_{aFt}^Q \equiv e^{-i\hat{H}_0^Q t/\hbar} |\psi_{aF}\rangle\langle\psi_{aF}| e^{i\hat{H}_0^Q t/\hbar}, \tag{4.65}$$

$$\hat{\xi}_{bIt}^Q \equiv e^{-i\hat{H}_0^Q t/\hbar} |\psi_b\rangle\langle\psi_b| e^{i\hat{H}_0^Q t/\hbar}, \tag{4.66}$$

$$\hat{\sigma}_{Ft}^D \equiv \mathcal{W} \cdot e^{-i\hat{H}_0^D t/\hbar} \hat{\rho}_F^D e^{i\hat{H}_0^D t/\hbar}, \tag{4.67}$$

$$\hat{\sigma}_{It}^D \equiv \mathcal{W} \cdot e^{-i\hat{H}_0^D t/\hbar} \hat{\rho}_I^D e^{i\hat{H}_0^D t/\hbar}. \tag{4.68}$$

Obviously, (4.63) is an explicit expression for the *wave function collapse*, characterized by the lack of the off-diagonal components, just like (3.5), as was to be expected.

Notice that the whole process is performed with one detector D_a placed in channel A, so that (4.63) exactly describes the WFC even in the case of *negative-result measurements*. We stress again that our theory has broken through the barrier of Wigner's theorem and has explicitly solved the negative-result-measurement paradox.

It is also interesting to rewrite (4.63) for the two types of detector, *non-destructive* and *destructive*. For an ideal non-destructive detector, (4.63) becomes

$$\hat{\Xi}_{It}^{tot} \xrightarrow{t \to \infty} \hat{\Xi}_{Ft}^{tot} = \overline{|T|^2} \hat{\xi}_{aFt}^Q \otimes \hat{\sigma}_{Ft}^D + \hat{\xi}_{bIt}^Q \otimes \hat{\sigma}_{It}^D + O(1 - \overline{|T|^2}) \tag{4.69}$$

with $\overline{|T|^2} \simeq 1$. On the other hand, for an ideal destructive detector, we have

$$\hat{\Xi}_{It}^{tot} \xrightarrow{t \to \infty} \hat{\Xi}_{Ft}^{tot} = (1 - \overline{|T|^2}) |0\rangle\langle 0| \otimes \hat{\sigma}_{Ft}^D + \hat{\xi}_{bIt}^Q \otimes \hat{\sigma}_{It}^D + O(\overline{|T|^2}) \tag{4.70}$$

with $\overline{|T|^2} \simeq 0$. Note that we have neglected the possibility of reflection at D_a, and have assumed that $\overline{|T'|^2} = 1 - \overline{|T|^2}$. We remark that all secondary processes, such as counter triggering, following the WFC, are described through the time evolution of $\hat{\sigma}_{Ft}^D$ in (4.69) and (4.70), or in general, in (4.63).

It is now worth discussing and clarifying the differences and analogies between a "detector" and a "dephaser." The above S-matrix analysis clearly shows that *detection implies dephasing*. However, the two concepts must be distinguished, because, as was explained before, the display process (which usually takes place at the last step of a measurement) can be conceptually separated from the former two steps (namely spectral decomposition and *bona fide* detection): The detector is simply acting as a dephaser during the second step. For this reason, in general, the word "dephaser" can be replaced by "detector," at least in the discussion on the dephasing process yielding the WFC. It goes without saying, however, that we cannot forget the other important function of the detector: its ability to produce signals, displaying the result of the measurement. This means that the above discussion is only valid in a restricted sense.

Moreover, and independently from the above discussion, we think that there exists no clear-cut distinction, at a physical level, between a measurement performed by a human observer and a natural process (where, by "natural," we mean something taking place in Nature, independently of the presence of sentient beings). This is a reflection of our philosophical position. The human will is only reflected in the design and in the way of setting up the whole apparatus accomplishing the above three steps, and the human activity is thus simply confined to the reading of the experimental results.

Finally, we would like to pay special attention to the detection of a single particle on a screen in the case of an interference experiment of the Young type, because many physicists used to consider satisfactory a description in terms of the naive WFC (1.7), relying upon the projection postulate. We are going to formulate the position measurement of a particle on a screen, within the framework of the MHS theory.

The particle position has a continuous spectrum, covering the whole screen. (Note that, contrary to this, the above-mentioned yes-no experiments only deal with the observation of a dichotomic observable by making use of two channels A and B.) We cannot *sharply* determine the particle position, as a mathematical point, by means of a physical process, but can only know in which (small) region the particle exists. For simplicity, let the position variable x be one-dimensional. If the distance between the fence (endowed with small slits) and the screen is sufficiently large, we can consider the particle wave function to be nearly constant on a macroscopically small region of the screen. The small region, which is very large on the microscopic scale, covers a "grain," where a spot will appear after the measurement. In this case we can decompose the continuous spectrum of x into a finite, nonoverlapping sequence of small regions I_i, centered around a central value x_i with volume Δ_i ($i = 1, 2, ..., N$), and then discretize the spectrum of the wave function in the following way

$$|\psi\rangle = \sum_i \psi_i |i\rangle, \quad \psi_i = \langle i|\psi\rangle, \tag{4.71}$$

where

$$|i\rangle \equiv \frac{1}{\sqrt{\Delta_i}} \int_{I_i} dx |x\rangle, \quad \langle i|j\rangle = \delta_{ij}. \tag{4.72}$$

Here the particle position can be described in terms of the discretized operator defined by the eigenvalue equation $\hat{X}|i\rangle = x_i|i\rangle$. Equation (4.71) represents the spectral decomposition with respect to the position measurement.

The total density matrix for the initial state just before the position measurement reads

$$\hat{\Xi}_I^{tot} = |\psi\rangle\langle\psi| \otimes \prod_k \hat{\sigma}_{Ik}^D = \sum_{ij} \psi_i \psi_j^* |i\rangle\langle j| \otimes \prod_k \hat{\sigma}_{Ik}^D, \tag{4.73}$$

where $\hat{\sigma}_{Ik}$ is the initial density matrix of the kth "grain" on the screen. Note that the screen can be thought of as covered with a huge number of small detectors ("grains"). In this density matrix the phase correlation among different $|i\rangle$s (or ψ_is) is fully kept. Needless to say, the position measurement will erase the phase correlation. Thus, the measurement process is written as

$$\hat{\Xi}_I^{tot} \longrightarrow \hat{\Xi}_F^{tot} = \sum_k |\psi_k|^2 \hat{\xi}_{Fk}^Q \otimes \hat{\sigma}_{Fk}^D \otimes \prod_{k' \neq k} \hat{\sigma}_{Ik'}^D, \tag{4.74}$$

where $\hat{\xi}_{Fk}^Q = |k\rangle\langle k|$ is the projection operator onto the position of the kth "grain," and $\hat{\sigma}_{Fk}^D$ the final density matrix of the kth "grain" after detection. In the limit of infinite number of "grains," we obtain a continuous distribution proportional to $|\psi(x)|^2$, as was already mentioned in (1.14).

2.3   General measurement process

The extension of the measurement theory outlined in the preceding subsection to the general case is straightforward. Consider the measurement, performed by detector D, of an observable $\hat{F}$ (its eigenvalues and eigenstates being λ_i and $|u_i\rangle$, respectively) on system Q, in the initial state $|\psi_0^Q\rangle = \sum_i c_i |u_i\rangle|\phi_0\rangle$, $|\phi_0\rangle$ being the wave packet. First of all, assume that we have realized the spectral decomposition (1.2), which is written in terms of the Q density matrix as follows:

$$\hat{\rho}_0^Q = |\psi_0^Q\rangle\langle\psi_0^Q| = \sum_{i,j} c_i c_j^* |\chi_i^{(0)}\rangle\langle\chi_j^{(0)}| \longrightarrow |\psi_I^Q\rangle\langle\psi_I^Q| = \sum_{i,j} c_i c_j^* |\chi_i\rangle\langle\chi_j| \equiv \hat{\rho}_I^Q, \tag{4.75}$$

where $|\chi_i^{(0)}\rangle \equiv |u_i\rangle|\phi_0\rangle$, $|\chi_i\rangle \equiv |v_i\rangle|\phi_i\rangle$ and $|\psi_I^Q\rangle = \sum_i c_i|\chi_i\rangle$. For the notation, see (1.2) and refer to Section 1.2, Chapter 1. Due to the spectral decomposition process, the ith branch wave corresponding to the ith eigenvalue of the observable $\hat{F}$ is forwarded to the ith channel I_i, which is spatially separated from other channels. The density matrix $\hat{\rho}_I^Q$ represents the initial Q state before the detection step; it is still a pure state, keeping the phase correlations among the branch waves.

The detection step will be performed by detector D_i, located in channel I_i. By observing the Q particle in channel I_i, one infers that the observable $\hat{F}$ of system Q has taken the ith eigenvalue. Denote the total number of channels by N_I. The total number of detectors is taken equal to N_I. Consequently, N_I collision processes take place, and are described by the S-matrix

$$\hat{S}^{(\ell)} = \prod_{i=1}^{N_I} \hat{S}_i^{(\ell)}, \tag{4.76}$$

in which each $\hat{S}_i$ describes an independent collision process in channel I_i. Having given the S-matrix, the measurement process is now expressed as

$$\hat{\Xi}_{It}^{tot} \overset{t\to\infty}{\longrightarrow} \hat{\Xi}_{Ft}^{tot} = \frac{1}{N_p} \sum_{\ell=1}^{N_p} e^{-i\hat{H}_0^{(\ell)}t/\hbar} \hat{S}^{(\ell)} \{\hat{\rho}_I^Q \otimes \hat{\rho}_I^{D(\ell)}\} \hat{S}^{(\ell)\dagger} e^{i\hat{H}_0^{(\ell)}t/\hbar}$$

$$= W \cdot e^{-i\hat{H}_0 t/\hbar} \hat{S} \{\hat{\rho}_I^Q \otimes \hat{\rho}_I^D\} \hat{S}^\dagger e^{i\hat{H}_0 t/\hbar}. \tag{4.77}$$

At the last step, we have used the ergodic assumption (4.25) to replace the event-average (bar-average) with the ensemble average. Our aim is to show that the final state is given by the expression

$$\hat{\Xi}_{Ft}^{tot} = \sum_k |c_k|^2 \hat{\xi}_{F(k)t}^Q \otimes \hat{\sigma}_{F(k)t}^D, \tag{4.78}$$

where $\hat{\sigma}_{F(k)t}^D$ stands for the final density matrix of the detector displaying the kth eigenvalue of $\hat{F}$ [see (3.5)].

The proof is straightforward: Set $W = \prod_i W_i$ and $\hat{\rho}_I^D = \otimes_i \hat{\rho}_I^{D_i}$ (corresponding to the presence of many detectors), $\hat{\rho}_I^{D_i}$ being the initial density matrix of detector D_i, placed in channel I_i and corresponding to the ith eigenvalue of $\hat{F}$. Write (4.77) as

$$\hat{\Xi}_{Ft}^{tot} = \sum_k |c_k|^2 \hat{\Xi}_{Ft}^{kk} + \sum_{k\neq k'} c_k c_{k'}^* \hat{\Xi}_{Ft}^{kk'}, \tag{4.79}$$

where the diagonal components are

$$\hat{\Xi}_{Ft}^{kk} = \prod_{i\neq k} W_i \cdot \left\{ W_k \cdot e^{-i\hat{H}_0 t/\hbar} \hat{S} \left[|k\rangle\langle k| \otimes \hat{\rho}_I^{D_k} \otimes \prod_{i\neq k} \hat{\rho}_I^{D_i} \right] \hat{S}^\dagger e^{i\hat{H}_0 t/\hbar} \right\} \tag{4.80}$$

and the off-diagonal components ($k \neq k'$) are

$$\hat{\Xi}_{Ft}^{kk'} = \prod_{s \neq k, k'} \mathcal{W}_s \cdot$$

$$\times \left\{ \mathcal{W}_k \cdot \mathcal{W}_{k'} \cdot e^{-i\hat{H}_0 t/\hbar} \hat{S} \left[|k\rangle\langle k'| \otimes \hat{\rho}_{I}^{D_k} \otimes \hat{\rho}_{I}^{D_{k'}} \otimes \prod_{s \neq k, k'} \hat{\rho}_{I}^{D_s} \right] \hat{S}^\dagger e^{i\hat{H}_0 t/\hbar} \right\}.$$

$$(4.81)$$

Notice that the S-matrix in (4.80) brings in an interaction between the states with a dot, yielding the phase factor $\exp(i\Theta_k)$ for the pair $(|k\rangle, \hat{\rho}_I^{D_k})$, and its conjugate $\exp(-i\Theta_k)$ for the pair $(\langle k|, \hat{\rho}_I^{D_k})$, if we take into account the structure (4.51) of the S-matrix and the discussion of Section 2.1. We can easily understand that these phase factors cancel out in the integral

$$\int d\ell_k W(\ell_k) e^{i\Theta_k} e^{-i\Theta_k}, \qquad (4.82)$$

so that the diagonal components do not vanish, even under the condition (4.46). By contrast, the S-matrix in (4.81) brings in interactions between the states with the dot or the circle, respectively, yielding different phase factors for the pairs $(|k\rangle, \hat{\rho}_I^{D_k})$ and $(\langle k'|, \hat{\rho}_I^{D_{k'}})$, which do not cancel out (because $k \neq k'$). Therefore we obtain integrals of the type

$$\int d\ell_k W(\ell_k) e^{i\Theta_k} \cdot \int d\ell_{k'} W(\ell_{k'}) e^{-i\Theta_{k'}}, \qquad (4.83)$$

which are vanishingly small under the condition (4.46).

In conclusion, we obtain the measurement process

$$\hat{\Xi}_{It}^{tot} \xrightarrow{t \to \infty} \hat{\Xi}_{Ft}^{tot} = \sum_k |c_k|^2 \hat{\xi}_{F(k)t}^{Q} \otimes \hat{\sigma}_{F(k)t}^{D_k} \otimes \prod_{i \neq k} \hat{\sigma}_{It}^{D_i}, \qquad (4.84)$$

where $\hat{\xi}_{F(k)t}^{Q}$ and $\hat{\sigma}_{F(k)t}^{D_k}$ are respectively the final density matrices of Q and D_k. Note that

$$\mathcal{W} \cdot \hat{S} \left[|k\rangle\langle k| \otimes \hat{\rho}_I^{D_k} \right] \hat{S}^\dagger = \hat{\xi}_{F(k)}^{Q} \otimes \hat{\sigma}_{F(k)}^{D_k}, \quad \hat{\sigma}_{F(k)}^{D} = \hat{\sigma}_{F(k)}^{D_k} \otimes \prod_{i \neq k} \hat{\sigma}_{Ii}^{D_i}. \qquad (4.85)$$

Once again, we need to comment on the fact that the exact wave function collapse can be realized only in the infinite N limit. This situation is somewhat similar to the one encountered in the theory of phase transitions, in which, rigorously speaking, the phase transition itself occurs only in the limit of infinite degrees of freedom. This implies that we can perform only *imperfect* measurements by means of measuring apparata with finite degrees of freedom. In this sense, the wave function collapse by measurement is a sort of asymptotic phenomenon similar to a phase transition. In practice, however, the infinite N limit is not strictly necessary in order to show the loss of quantum mechanical coherence (i.e. the wave function collapse): This will be shown in Chapter 7 by means of a numerical simulation of the measurement process, and is reminiscent of many numerical simulations of phase transitions. In this context, we stress that the wave function collapse can be practically realized by means of measuring apparata with sufficiently large N.

On the other hand, we remark that it is possible to conceive a macroscopic instrument which is endowed with a very large N but does not necessarily provoke a loss of quantum mechanical coherence: A good example is a phase shifter in neutron interferometry. This situation can occur if $\sum_{n=1}^{N} \kappa_n a_n / \kappa_0$ in (4.24) and $N\overline{\kappa^2}(\Delta a)^2$ in (4.33) are finite (for simplicity, we are suppressing the suffix labelling the detector), even though N goes to infinity, so that the Riemann-Lebesgue lemma cannot be applied to the statistical average. In this case, it may be convenient to introduce a scale transformation from the microscopic variable $\sum_{n=1}^{N} \kappa_n a_n / \kappa_0$, with discrete N, to a continuous macroscopic variable (like a length) in the infinite N limit, as was done in Section *2.1*.

3   Observation of a single system

Since the MHS theory has introduced many Hilbert spaces in order to describe the measurement process performed on many Q and D (local) systems, one can ask: Is the MHS theory able to deal with the observation of a single system? The answer is "yes" in the sense explained below.

In order to tackle this problem, let us first consider an interference experiment of the Young type, in which we observe the wave-particle nature of a quantum mechanical "particle" on the accumulated distribution of many independent spots produced by many incoming particles (a particle beam) sent into the apparatus. Notice that, on the other hand, if the wave-particle nature of the incoming particle is well understood in advance, the experiment is performed with a different purpose: one may want to obtain significant information on the structure of the slit system. In this case, the slit itself is a single macroscopic system, but in order to know its structure, we need to use a huge number of incoming particles (say, photons) and observe the pattern on the screen, as explained in Section *2.2*. We endeavour to describe this process within the framework of the MHS theory. Notice that even though the roles played by Q and D are interchanged here, the elementary interaction between them keeps the same form and meaning as in the previous cases considered and we are still interested in the effect of fluctuations in D (now *a measured object*) on Q (*measuring object*).

Of course, the slit system just introduced is only schematic and ideal; in reality, the slit system is a crystal. In this sense the experiment can be viewed as an experimental analysis of the crystal structure. In two interesting experiments, Walther and his group[167] and Itano *et al.*[86] replaced the slit system (the crystal) with a single atom, yet used a light beam, composed of many photons, and accumulated many single events, in order to obtain a picture of the single atom. The MHS theory can also deal with such an experiment without essential modifications. In this case, the expected statistical fluctuations become extremely small for the accumulated distribution since the single atomic system is considered to be in a well-defined microscopic state.

The statistical nature of the MHS theory is based on the fact that quantum mechanics can never give a definite prediction for a single measurement on a single system, but can only predict a definite result for the accumulated distribution given by many independent events performed on many dynamical systems, each of which is described by the same wave function. However, this fundamental nature of quantum mechanics does not exclude the special case in which we have a definite result in a single experiment when each dynamical system is in an eigenstate of the observable to be measured. In this case, since after the spectral decomposition only one branch wave, which is in an eigenstate belonging to the eigenvalue to be measured, exists, we can see no interference at all, irrespectively of the state of the measuring system. We need not repeat the experiment and accumulate the results of many experiments, as far as we know that the object is in an eigenstate of the observable to be measured. Usually, however, we do not know whether Q is in an eigenstate of the measured observable. In this case, we have to repeat the experiment and accumulate many results, in order to infer that the system is in an eigenstate of the observable.

In addition to this, we have another reason to accumulate many experimental results, in order to make an image of a single system such as, for example, an atom. In the case of the observation of a single system, the state of the object is in an eigenstate of the number of atoms belonging to the eigenvalue 1, but not in a simultaneous eigenstate of other observables, such as electron positions inside the atom. Therefore, we should observe a statistical fluctuation of electron positions in an experimental run, which will produce a diffused image of the atom for the accumulated distribution of many experimental results. This procedure is nothing but the MHS-theoretical one.

This kind of situation is to be compared with the observation of a classical system, for which it is sometimes believed that one need not provide an ensemble of many classical systems, because one is dealing with a single object system. In such a case, we can compare the classical system with the above single atom, and make use of light composed of many photons in order to analyze its structure. Moreover, we should pay attention to the fact that a classical system is represented by a few macroscopic dynamical observables, for which the quantum mechanical restrictions arising from the uncertainty principle can be neglected, on a macroscopic scale. In other words, the system is considered (with sufficient accuracy on a macroscopic scale) to be approximately in a simultaneous eigenstate of the dynamical observables. Thus, the statistical fluctuations of the accumulated distribution become very small, on a macroscopic scale. Even in this case, fluctuations on a microscopic scale do exist for the classical system and they may cause the reduction of the quantum mechanical coherence of the particle if the condition for decoherence, i.e. (4.46) or (4.35), is met. As for the observation of a single system, however, no essential difference from the quantum case mentioned above can be found in this case.

There have been many attempts, in the past, to circumvent the impossibility of describing the outcomes of single events, thereby explaining what we called "naive" wave function collapse (see, for example, Refs. 26, 67, 156). In these cases, the authors endeavour to describe *individual quantum mechanical events*. Observe that these theories go beyond quantum mechanics, either by modifying it in some essential way, or by "completing" it via the introduction of (nonlocal) hidden parameters. In any case, these approaches lie outside quantum mechanics.

We stress that our point of view is different. The MHS approach never attempts to describe individual events, but rather endeavours to derive the collapse of the wave function for the accumulation of many single events. The collapse, and more generally the loss of quantum coherence, is obtained only for the accumulated distribution of many individual events. Note, however, that (3.5) itself expresses a sum of probabilities of mutually exclusive events, which implies that once one event occurs, all other events never take place. This is our notion of *wave function collapse*.

4   The EPR problem in the MHS theory

The EPR problem was briefly introduced in Section 4 of Chapter 3. It is interesting to comment on this problem from the MHS point of view, which forces us to consider the two-step structure (spectral decomposition and detection) of a measurement process and the presence of the apparatus states. Suppose that the initial EPR state is given by

$$\psi_0^Q = |\chi\rangle\Phi_0(r_1, r_2; t), \tag{4.86}$$

$$|\chi\rangle = \frac{1}{\sqrt{2}}[|+z\rangle_1|-z\rangle_2 - |-z\rangle_1|+z\rangle_2],$$

where $|\chi\rangle$ is the singlet state [refer to (3.35) for the notation] and Φ_0 the initial wave packet of the two particles, moving in space after the correlation described by $|\chi\rangle$ is established. Notice that Q stands for *two* quantum particles, now. Let us first perform a measurement of the z-spin component of particle #1, as in the discussion following Eq. (3.35): assume that the wave packets of the two particles are well separated after a certain time t_0 and write $\Phi_0(r_1, r_2; t) = \phi_1(r_1, t)\phi_2(r_2, t)$ $(t > t_0)$, where $\phi_i(r_i, t)$ is the wave packet of particle #i. At time $t_1 > t_0$ the wave packet of particle #1 meets the spectral decomposer, set up to prepare a measurement of $\sigma_z^{(1)}$. The decomposer will perform the spectral decomposition

$$\psi^Q = \frac{1}{\sqrt{2}}\left[|+z\rangle_1|-z\rangle_2\phi_+(r_1, t)\phi_2(r_2, t) - |-z\rangle_1|+z\rangle_2\phi_-(r_1, t)\phi_2(r_2, t)\right],$$

(4.87)

where $\phi_\pm$ is the wave packet of particle #1 running in channel $\pm$. Channels $\pm$ are separated with respect to r_1. We place a detector D in channel +, but no detector in channel −. In the naive description (1.7) of the WFC, the spin measurement will yield

$$\psi^Q \longrightarrow |+z\rangle_1|-z\rangle_2\phi_+(r_1, t)\phi_2(r_2, t) \quad \text{or} \quad |-z\rangle_1|+z\rangle_2\phi_-(r_1, t)\phi_2(r_2, t)$$

(4.88)

if we observe the value +1 or −1 for $\sigma_z^{(1)}$, respectively. We can draw the same conclusions as in Section 4 of Chapter 3.

As repeatedly emphasized in this book, however, the measurement process should be described in terms of density matrices. The measurement yields

$$\Xi_I^{tot} \longrightarrow \Xi_{Ft}^{tot} = \rho_{Ft}^{Q+} \otimes \sigma_{Ft}^D + \rho_{Ft}^{Q-} \otimes \sigma_{It}^D$$

(4.89)

where Ξ is the density matrix of the total system,

$$\rho_{Ft}^{Q\pm} = |\pm z\rangle_{11}\langle \pm z| \otimes |\mp z\rangle_{22}\langle \mp z| \cdot |\phi_\pm(r_1, t)|^2|\phi_2(r_2, t)|^2$$

(4.90)

is the density matrix of the two Q particles and $\sigma_{Ft}^D, \sigma_{It}^D$ are the detector states displaying detection and no detection, respectively. The lack of cross-correlated terms in (4.89) means that the wave function collapse has taken place.

Equation (4.89) simply expresses the loss of coherence (WFC) and does not really add new insight into the EPR phenomenon. On the other hand, consider a *different* experiment, devised to perform a measurement of the x-spin component of particle #1, as in the discussion following Eq. (3.36). The singlet state is conveniently written as in (3.36):

$$|\chi\rangle = \frac{1}{\sqrt{2}}\left[|+x\rangle_1|-x\rangle_2 - |-x\rangle_1|+x\rangle_2\right]$$

(4.91)

and at time $t_1 > t_0$ the wave packet of particle #1 meets a *different* spectral decomposer, set up to prepare a measurement of $\sigma_x^{(1)}$. The corresponding spectral decomposition is

$$\psi'^Q = \frac{1}{\sqrt{2}} \left[| + x\rangle_1| - x\rangle_2 \phi'_+(\mathbf{r}_1, t)\phi_2(\mathbf{r}_2, t) - | - x\rangle_1| + x\rangle_2 \phi'_-(\mathbf{r}_1, t)\phi_2(\mathbf{r}_2, t) \right],$$

$$(4.92)$$

where $\phi'_{\pm}$ is the wave packet of particle #1 running in the new channels $\pm$, pertaining to the x-component of the spin, in this case. The new measurement, performed by detector D′ placed in the new channel +, yields [in the notation of (4.89)]

$$\Xi'^{tot}_I \longrightarrow \Xi'^{tot}_{Ft} = \rho'^{Q+}_{Ft} \otimes \sigma^{D'}_{Ft} + \rho'^{Q-}_{Ft} \otimes \sigma^{D'}_{It}$$

$$(4.93)$$

with

$$\rho'^{Q\pm}_{Ft} = | \pm x\rangle_{11}\langle \pm x| \otimes | \mp x\rangle_{22}\langle \mp x| \cdot |\phi'_{\pm}(\mathbf{r}_1, t)|^2 |\phi_2(\mathbf{r}_2, t)|^2.$$

$$(4.94)$$

We stress that even the initial density matrix Ξ'^{tot}_I is different, because it describes a *different* initial experimental setup (spectral decomposer aligned in another direction, as well as a detector placed at a different position in space).

Although the above conclusion is a straightforward consequence of a rigorous application of quantum mechanics to the total Q+D system (as a matter of fact, the MHS approach forces us to follow faithfully the actual realization of an experiment), the difference between Eqs. (4.89)-(4.90) and (4.93)-(4.94) is noteworthy. It is obvious that any spectral decomposition, performed along a different direction u, would yield a different final density matrix for the total system. As we shall see, this leads to interesting consequences, from the MHS point of view.

Let us make this point clear with a simple example. By following the spirit of the approach outlined in the present book, one identifies the event-average ($\overline{\cdots}$), with the average over the space of the variables describing the whole apparatus (spectral decomposer + detector). On the other hand, an advocate of local hidden variable theories would naturally be lead to identify

$$\overline{\cdots} \equiv \int_\Lambda d\mu(\lambda) \cdots, \qquad (4.95)$$

where λ, introduced in (3.37) and (3.38), is the (set of) local hidden variable(s), distributed over the space Λ with probability measure $\mu(\lambda)$. It was stressed after Eq. (3.38) that Λ and μ neither depend on a, nor on b, and $A(a, \lambda)$, $B(b, \lambda)$ do not depend on b, a, respectively. This brings up an interesting problem: According to (4.25) and (4.31), we have

$$\int_\Lambda d\mu(\lambda) \cdots = \overline{\cdots} = \langle \cdots \rangle = \mathcal{W} \cdot \{\cdots\}, \qquad (4.96)$$

so that the λ-average, characterizing the particles in the source, is reflected in the $\mathcal{W}$-average, characterizing the whole detection system.[d] When written in terms of Bell's observable (3.38), that involves a measurement on two correlated particles, performed by two distinct detection apparata, the above condition reads

$$\int_\Lambda d\mu(\lambda) \cdots = \mathcal{W}_a \cdot \mathcal{W}_b \cdot \{\cdots\}, \tag{4.97}$$

where we have explicitly written the dependence of the $\mathcal{W}$-averages on the macroscopic parameters characterizing the detection system, composed of, say, detectors D_a and D_b. The r.h.s. of (4.97) acts like in (4.80)-(4.81). (Notice that we are *formally* including the Stern-Gerlach magnet in the detection system. Obviously, this does not mean that coherence is destroyed by the spectral decomposition.) On the other hand, in order to check Bell's inequality (3.39), other experiments must also be performed, with different orientations $(a, b'), (a', b), (a', b')$. However, Bell's inequality (3.39) cannot be derived, unless we require

$$\mathcal{W}_a \cdot \mathcal{W}_a \cdot \equiv \mathcal{W}_{a'} \cdot \mathcal{W}_b \cdot \equiv \mathcal{W}_a \cdot \mathcal{W}_{b'} \cdot \equiv \mathcal{W}_{a'} \cdot \mathcal{W}_{b'} \cdot. \tag{4.98}$$

Can the assumption (4.98) be considered "natural"? This is a delicate problem, that leads us to the so-called counterfactuality argument, proposed by several authors in a different context,[179] in order to follow Bohr's philosophy as closely as possible. Observe that if we formalize the detection process according to the MHS theory, the density matrix of the apparata must be considered *ab initio* in the calculation. On the other hand, it is impossible, even in principle, to compute the evolution of the total system when the Stern-Gerlach magnets and the detection system correspond to two mutually incompatible observables, such as a and a' or b and b' of the previous example. In this sense, no *concrete* physical situation can correspond to the requirement (4.98). Notice that it is fundamental for the derivation of the inequality (3.39).

We shall not discuss here the consequences of the assumption (4.98), and shall just observe that, from a strictly operationalistic standpoint, *counterfactuality* itself is a rather vague concept. In this sense, the above discussion must be regarded only as an attempt at understanding which hypotheses are necessary in order to derive the inequality (3.39).

It should also be stressed that the above discussion holds true for any initial state of the object particles. In particular, when discussing the validity of Eq. (4.98), no reference is made to the state $|\chi\rangle$. It would seem, therefore, that if (4.98) were true, the nonlocal aspects of quantum mechanics stem from

---

[d]Equation (4.96) is by no means obvious, because hidden variables are very different from macroscopic fluctuating variables, and any "ergodic-like" assumption would require justification, for the former.

the detection process, rather than from the existence of entangled[175] quantum mechanical states, such as $|\chi\rangle$. This is in contrast with the commonly accepted view on the meaning of the Bell inequality.

The problem of quantum nonlocality certainly requires a deeper analysis than the present one. It is interesting to observe, though, that an MHS formalization of the EPR problem is far from being trivial, and seems to point out a delicate aspect of measurements performed on correlated two-particle states.

5   Quantum non-demolition measurements

As was repeatedly mentioned, a quantum measurement provokes a remarkable transition of the object system from the initial superposed state to the final mixed state through a dephasing process among the branch waves. This means that a quantum measurement yields, in general, a dramatic change of the initial quantum state. In particular, as was discussed in Section 2.2, the largest modification of the initial state occurs in the case of photon and neutron detection, because these particles are usually detected by making use of a *destructive* detector, which absorbs or destroys them. In other words, the particle does not exist anymore after the measurement and the difference between the states before and after the measurement is enormous. Recently, however, even photons can be detected via a so-called *quantum non-demolition* measurement, without resorting to any destructive process. In this section, we shall discuss quantum non-demolition measurements from the MHS-theoretical point of view.

The original idea of the quantum non-demolition measurement is rooted in the discussion concerning the detection of gravitational waves coming from some remote source in the Universe. It is well known that, under normal conditions, the signals of such gravitational waves cannot be detected, because they are small enough to be masked by the quantum zero-point fluctuations (i.e. vacuum fluctuations, see discussion in Section 5 of Chapter 2). Braginsky[31,32] proposed to the use of a stroboscopic measurement for the observation of an oscillator coordinate driven by gravitational waves: In such a scheme, called *quantum non-demolition measurement* (QND), there is a device resonating at the zeros of the different-time commutator of the Heisenberg position operator. Such a measurement, if feasible, should overcome the problems of the uncertainty relations arising from the zero-point fluctuations, and yield a definite result. Braginsky and others succeeded in formulating the general conditions for a quantum non-demolition measurement. In principle, such a quantum non-demolition measurement should be feasible, but unfortunately, in many cases, its experimental realization is not easy.

At present, we can find practical examples of quantum non-demolition measurements only in the field of quantum optics,[116,197,119] in the case of photon detection by means of strong laser beams. In such a case, we have to prepare a probe and a signal beam in different modes,[e] in order to infer the quantum state of the former by measuring the state of the latter. To this end, both beams must be strongly correlated via, for example, a nonlinear interaction. Usually, in optical phenomena, such nonlinear effects are so small that one has to use strong laser beams in order to make the responses large enough. This is the reason why quantum non-demolition measurements cannot be easily realized in the case of weak signal beams.

Here we only outline a typical example of a quantum non-demolition measurement of the photon state of a signal beam, by making use of the Kerr effect. For more details and similar ideas, refer to Ref. 116, 197, 119. When a medium is crossed by a strong light beam, its refractive index suffers a change proportional to the light intensity of the beam. Remember that we are dealing with two modes corresponding to a probe and a signal, respectively. Such a change of the refractive index is produced not only by a given mode on its own refractive index, but also by a given mode on the refractive index of the other mode. For a very weak intensity, it is enough to deal with the linear interaction of mode μ given by the Hamiltonian

$$\hat{H}_1 = \kappa_1 \hat{a}_\mu^\dagger \hat{a}_\mu, \tag{4.99}$$

where κ_1 is a constant proportional to the laser electric field concerned. However, in the situation just described, we have to take into account the third order non-linear interaction, described by the effective Hamiltonian

$$\hat{H}_3 = \kappa_3 \hat{n}_\mu \hat{n}_\nu, \tag{4.100}$$

where $\hat{n}_\mu = \hat{a}_\mu^\dagger \hat{a}_\mu$, $\hat{n}_\nu = \hat{a}_\nu^\dagger \hat{a}_\nu$ and κ_3 is a constant proportional to the square of the bilinear term of the laser electric field concerned. Identify μ and ν with the probe and signal mode, respectively. The time evolution operator due to $\hat{H}_3$ is given by

$$\hat{U}(T) = e^{-iT\kappa_3 \hat{n}_\mu \hat{n}_\nu / \hbar}, \tag{4.101}$$

T being the common passage time of the probe and signal waves in the middle channel in Fig. 4.2, so that both operators evolve according to

$$\hat{a}_\mu(T) = \hat{U}^\dagger(T) \hat{a}_\mu \hat{U}(T) = e^{-iT\kappa_3 \hat{n}_\nu / \hbar} \hat{a}_\mu,$$
$$\hat{a}_\nu(T) = \hat{U}^\dagger(T) \hat{a}_\nu \hat{U}(T) = e^{-iT\kappa_3 \hat{n}_\mu / \hbar} \hat{a}_\nu. \tag{4.102}$$

Look at the first relation: It implies that we can infer the signal (mode ν) number state by observing the phase change of the probe (mode μ) beam, i.e. by analyzing the interference at D_0 of the probe with its own reference beam crossing the upper channel. This observation can be performed without destroying

---

[e] As for "mode," see the discussion in Section 5 of Chapter 2.

the signal number state. This is nothing but a quantum non-demolition measurement.

In order to better comprehend such a measurement, it is convenient to think of the whole experiment as divided in the following steps: At the first step, two (strong) incoming laser beams, P (mode μ) and S (mode ν), corresponding to the probe and the signal, come into the instrument from the left, and are then divided by the beam splitters V_P and V_S, into $P \rightarrow P_a + P_c$ and $S \rightarrow S_a + S_b$, respectively. At the second step, beams P_a and S_a are combined in the middle channel A and then divided by the beam splitter V_A into P'_a and S'_a, respectively. At the third step, beams P_c and P'_a are combined in the final upper channel, while beams S_b and S'_a are combined in the final lower channel. Suppose that, due to a strong laser electric field, the Kerr effect takes place

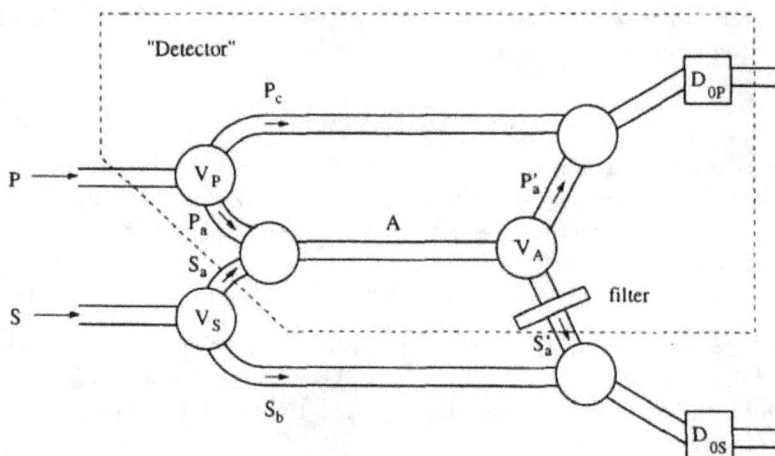

Fig. 4.2. An example of QND.

in the middle channel and is described by an interaction Hamiltonian (4.100), which yields the time evolution (4.102). As a final step, we perform an interference experiment at D_{OP} for the probe beam, by which we infer the signal number state through the displacement of the interference fringes. [Remember that the P-phase shift is proportional to the S-number. See (4.102).] This kind of measurement has actually been carried out.

However, its feasibility is not so clear in the case of very weak signal beams, and has not been achieved at the time of writing. Nevertheless, there is an alternative possibility: Equation (4.102) holds still true if only the intensity of the probe beam is strong enough. In this case, we can expect to detect a small phase change of the probe beam, caused by the signal beam, at D_{0P} in the upper final channel. This is a genuine quantum non-demolition measurement for a very weak beam. In this case, however, we expect to have dephasing between the two branch signal waves, S'_a and S_b, crossing the two lower channels. Such a dephasing would cause a reduction of interference at D_{0S} in the lower final channel after recombination. In this case, we have to take into account the fact that, for example, the signal frequency is much higher than the probe frequency. This situation should imply that, even though the relative phase between the two probe waves P_c and P'_a, in the two upper channels, is sharply shifted because its fluctuations can be neglected, the phase shift between the two signal waves S_b and S'_a in the lower channels becomes random and therefore not sharply definite. Note that D_{0S} is preceded by a high-pass filter, through which waves of mode ν can pass but waves of mode μ cannot. The dephasing of the two branch waves in the lower channels is just what we have discussed in this book from the MHS-theoretical point of view.

At any rate, as far as we know, nobody has yet succeeded in realizing a genuine quantum non-demolition measurement. Nevertheless, we are confident that recent remarkable innovation in sophisticated techniques will make possible ideal quantum non-demolition measurements in the very near future.

Chapter 5

SOLVABLE DETECTOR MODELS

Let us analyze the mechanism at the origin of the loss of quantum mechanical coherence by discussing some solvable models. We shall see that coherence can be either completely or partially lost, depending on the physical conditions specifying the state of the measuring apparatus. This will also elucidate the concept of imperfect measurement. The models we are going to investigate are simple enough to be solvable, yet they provide interesting physical insight into the MHS approach. All models clarify one or more aspects of a quantum measurement.

1 Rigid mirror model

The first solvable model we shall consider is a perfect mirror with a rigid wall: it can be used to measure the momentum of a very energetic particle. The idea is the following:[117] If the incoming particle has got enough energy to displace the mirror by a macroscopic distance, then, by measuring the recoil momentum of the latter, we can infer the particle momentum before and after the collision. The above is obviously a *gedanken experiment*, even though one should notice that very high-energy cosmic rays can have enough energy to accomplish such a process, under the hypothesis that we can construct a perfectly rigid mirror.

The mirror is mounted on a small wagon, free to move on a horizontal rail. A small friction force, proportional to the velocity, acts between the wagon and the rail, and the collision can be considered elastic. Under these conditions, the wagon travels a distance $L = P'/f$, where f is the friction coefficient and P' the wagon momentum after the collision. We shall assume that the wagon mass M is much larger than the particle mass m, so that if p and p' are the particle momenta before and after the collision, respectively, one has $p' = -p$ and $P' = 2p$. The particle momentum can therefore be inferred by the formula $p = Lf/2$. The problem is essentially one dimensional.

Observe that the irreversible process caused by the above-mentioned friction force is introduced only with the purpose of displaying the result of the measurement and is not essential to derive the wave function collapse. One can perform the measurement by means of a purely dynamical process.

The center of mass of the mirror-wagon system can be expressed as a well localized wave packet $|P, X_0\rangle$, whose uncertainties δP and δX_0 around P and X_0 are very small on a macroscopic scale. According to the MHS approach, the center of mass of the mirror-wagon system is a macroscopic variable, that cannot be exactly determined on a microscopic scale, and must be described by the following density matrix[a]

$$\rho_0^{AX}(P, X) = \int dX_0 \, |P, X_0\rangle W^X(X - X_0)\langle P, X_0|, \tag{5.1}$$

where $W^X(X - X_0)$ is a normalized weight function with width δX around $X_0 = X$ and the superscript A (apparatus) denotes the mirror-wagon system.

We shall neglect the spread of the wave packet during the measurement process, so that the evolution of the system is given by

$$\rho_t^{AX}(P, X) = e^{-iH^{AX}t/\hbar}\rho_0^{AX}(P, X)e^{iH^{AX}t/\hbar} = \rho_0^{AX}\left(P, X + \frac{P}{M}t\right), \tag{5.2}$$

where H^{AX} is the free Hamiltonian for the motion of the center of mass and M the mass of the system. We should not forget, however, that the mirror inner state itself must be treated quantum mechanically. Let H^{AI} be the inner-state Hamiltonian, for which the eigenvalue equation reads

$$H^{AI}|n, D\rangle = E_n^{AI}|n, D\rangle, \tag{5.3}$$

where D is the mirror thickness. Once again, since the mirror is a macroscopic body, its thickness cannot be precisely determined on a microscopic scale, and we are led to the following representation for the inner state

$$\sigma_0^{AI}(d) = \int dD \, W^I(D - d)\rho_0^{AI}(D), \tag{5.4}$$

where $W^I(D - d)$ is a normalized weight function with width δD, centered around $D = d$, and

$$\rho_0^{AI}(D) = \sum_n w_n \xi_n^I(D),$$

$$\xi_n^I(D) = |n, D\rangle\langle n, D|, \tag{5.5}$$

---

[a] In what follows, for simplicity, the caret on operators will be suppressed unless confusion may arise.

with w_n given, for instance, by the Boltzmann distribution at thermal equilibrium. The quantity d is nothing but the mirror thickness in a macroscopic sense. Notice that the evolution is

$$\sigma_t^{AI}(d) = e^{-iH^{AI}t/\hbar}\sigma_0^{AI}(d)e^{iH^{AI}t/\hbar} = \sigma_0^{AI}(d), \qquad (5.6)$$

so that the inner state does not change. The statistical fluctuations of the mirror system stem only from the distribution of the thickness D around d with width δD. In this case, $W^I(D-d)$ is representative of the inner fluctuations of the detector.

The total state of the mirror-wagon system is given by the density matrix

$$\begin{aligned}\rho_0^A(P,X,d) &= \rho_0^{AX}(P,X)\otimes\sigma_0^{AI}(d)\\ &= \int dD\, W^I(D-d)\rho_0^A(P,X,D)\end{aligned} \qquad (5.7)$$

with $\rho_0^A(P,X,D) = \rho_0^{AX}(P,X)\otimes\rho_0^{AI}(D)$ and its evolution is

$$\rho_t^A(P,X,d) = e^{-iH^A t/\hbar}\rho_0^A(P,X,d)e^{iH^A t/\hbar} = \rho_0^A\left(P,X+\frac{P}{M}t,d\right), \qquad (5.8)$$

where $H^A = H^{AX} + H^{AI}$ is the total Hamiltonian of the mirror-wagon system. This is our "detector:" It is described by the macroscopic variables P, X and d. We shall take $\rho_0^A(0,0,d)$ as the initial state, and assume δD, δX and δX_0 macroscopically small but microscopically large, for instance larger than a typical atomic size. A possible choice, that underestimates δD and overestimates δP, is

$$\delta D \sim 10^{-8}\,\text{cm}, \qquad \delta P = \hbar/\delta D \sim 1\,\text{keV}/c. \qquad (5.9)$$

Notice the analogy between Eq. (4.43) and Eqs. (5.1), (5.4) and (5.7).

The state of the incoming particle, whose momentum we want to measure, is assumed to be a wave packet $|p,x_0\rangle$, of average momentum p, centered around x_0, with spread δx. Its momentum spread $\delta p = \hbar/2\delta x$ can be made arbitrarily small, so that we can assume

$$\delta P \gg \delta p \qquad \text{or equivalently} \qquad \delta x \gg \delta D. \qquad (5.10)$$

It should be kept in mind, however, that δx is very small on a macroscopic scale, so that $\delta x \ll d, L$. If we neglect the spread of the wave packet, the particle evolution is described by

$$\xi_t^Q(p,x_0) = e^{-iH^Q t/\hbar}\xi_0^Q(p,x_0)e^{iH^Q t/\hbar} = \xi_0^Q\left(p,x_0+\frac{p}{m}t\right), \qquad (5.11)$$

where $\xi_t^Q(p,x_0) = |p,x_0\rangle\langle p,x_0|$, and H^Q is the free particle Hamiltonian.

We are now ready to describe the measurement process. For simplicity, let the initial particle state be

$$|\psi\rangle = c_1|p_1, x_0\rangle + c_2|p_2, x_0\rangle, \tag{5.12}$$

or equivalently

$$\begin{aligned}
\rho_0^Q &= |\psi\rangle\langle\psi| \\
&= |c_1|^2\xi_0^Q(p_1, x_0) + |c_2|^2\xi_0^Q(p_2, x_0) \\
&\quad + c_1 c_2^* \eta_0^Q(p_1, p_2, x_0) + c_1^* c_2 \eta_0^Q(p_2, p_1, x_0),
\end{aligned} \tag{5.13}$$

where $\eta_0^Q(p_1, p_2, x_0) = |p_1, x_0\rangle\langle p_2, x_0|$. The extension to a general superposed state is straightforward.

The measurement process is described by

$$\rho_t = e^{-iHt/\hbar}\rho_0 e^{iHt/\hbar}, \tag{5.14}$$

where $\rho_0 = \rho_0^Q \otimes \rho_0^A(0, 0, d)$ is the initial state of the total system, and $H = H_0 + H_{\text{int}}$ the total Hamiltonian, written in terms of the free part $H_0 = H^Q + H^A$ and the interaction H_{int}. By Eq. (5.13), we can write

$$\rho_t = |c_1|^2\Xi_t^{11} + |c_2|^2\Xi_t^{22} + c_1 c_2^*\Xi_t^{12} + c_1^* c_2\Xi_t^{21} \tag{5.15}$$

where

$$\begin{aligned}
\Xi_t^{11} &= e^{-iHt/\hbar}\Xi_0^{11}e^{iHt/\hbar}, \\
\Xi_0^{11} &= \xi_0^Q(p_1, x_0) \otimes \rho_0^A(0, 0, d), \\
\Xi_t^{12} &= e^{-iHt/\hbar}\Xi_0^{12}e^{iHt/\hbar}, \\
\Xi_0^{12} &= \eta_0^Q(p_1, p_2, x_0) \otimes \rho_0^A(0, 0, d),
\end{aligned} \tag{5.16}$$

and analogously for the other terms. In the interaction picture, the evolution is governed by the S-matrix, defined as usual by

$$U_I(t) = e^{iH_0t/\hbar}e^{-iHt/\hbar} \xrightarrow{t \to \infty} S. \tag{5.17}$$

In the case here considered, the scattering between the particle and the mirror-wagon system is asymptotically equivalent to the scattering between the particle and a rigid wall placed at a distance $x = -D/2$ in the relative coordinate x, so that the S-matrix element is easily evaluated as

$$\langle p', P'|S|p, P\rangle = -e^{-ipD/\hbar}, \tag{5.18}$$

where, due to our initial assumptions, we have $P = 0, P' = 2p$ and $p' = -p$. Rigorously speaking, we can state that the phase shift is proportional to the product pD in the $p \to \infty$ limit. A straightforward calculation, based on the definition (5.17) and on the assumption (5.10), yields

$$
\Xi_t^{11} \overset{t\to\infty}{\longrightarrow} \xi_t^Q(-p_1, -x_0 - d) \otimes \sum_n \iint dX_0 dD
$$
$$
\times \left[-e^{-ip_1(D-d)/\hbar}\right]|2p_1, X_0\rangle_t W^X(X_0 - 0)\,_t\langle 2p_1, X_0|\left[-e^{+ip_1(D-d)/\hbar}\right]
$$
$$
\otimes|n, D\rangle w_n W^I(D - d)\langle n, D|
$$
$$
= \xi_t^Q(-p_1, -x_0 - d) \otimes \rho_0^A\left(2p_1, \frac{2p_1}{M}t, d\right), \tag{5.19}
$$

and analogously for Ξ_t^{22}, because the two phase factors cancel out. On the other hand,

$$
\Xi_t^{12} \overset{t\to\infty}{\longrightarrow} \eta_t^Q(-p_1, -p_2, -x_0 - d) \otimes \sum_n \iint dX_0 dD
$$
$$
\times \left[-e^{-ip_1(D-d)/\hbar}\right]|2p_1, X_0\rangle_t W^X(X_0 - 0)\,_t\langle 2p_2, X_0|\left[-e^{+ip_2(D-d)/\hbar}\right]
$$
$$
\otimes|n, D\rangle w_n W^I(D - d)\langle n, D|, \tag{5.20}
$$

and we see that the two phase factors $e^{-ip_1(D-d)/\hbar}$ and $e^{+ip_2(D-d)/\hbar}$ do not cancel out for $p_1 \neq p_2$: The off-diagonal term Ξ_t^{12}, in Eq. (5.20), therefore contains the integral

$$
\int dD\, W^I(D - d)e^{-i(p_1 - p_2)(D-d)/\hbar} \rho_0^{AI}(D). \tag{5.21}
$$

Since $d \gg \delta D$ and $\rho_0^{AI}(D)$ is a slowly varying function of D, the above integral vanishes if

$$
|p_1 - p_2| \gg \delta P, \tag{5.22}
$$

where δP is defined in Eq. (5.9). We obtain

$$
\Xi_t^{12} \overset{t\to\infty}{\longrightarrow} 0, \tag{5.23}
$$

and therefore

$$
\rho_t \overset{t\to\infty}{\longrightarrow} \rho_F = \Xi_t^{11} + \Xi_t^{22}
$$
$$
= |c_1|^2 \xi_t^Q(-p_1, -x_0 - d) \otimes \rho_0^A\left(2p_1, \frac{2p_1}{M}t, d\right)
$$
$$
+ |c_2|^2 \xi_t^Q(-p_2, -x_0 - d) \otimes \rho_0^A\left(2p_2, \frac{2p_2}{M}t, d\right), \tag{5.24}
$$

which is the explicit expression for the wave function collapse, characterized by the lack of off-diagonal terms. Note that this detector, under the condition (5.22), enables us to perform a momentum measurement without resorting to the spectral decomposition.

The wave function collapse is obtained in the present model if

$$|p_1 - p_2|\delta D \gg \hbar, \tag{5.25}$$

as can be inferred by Eq. (5.22). However, we stress that the condition (5.25) is not essential, in order to obtain the wave function collapse, and can be dispensed with by simply introducing a momentum analyzer in our model, in order to perform a spectral decomposition before the actual measurement, in a way similar to Fig. 1.2. Indeed, if we place mirror detectors in Fig. 1.2, the wave function collapse is obtained if

$$|p_i|\delta D \gg \hbar, \qquad i = 1, 2, \ldots \tag{5.26}$$

The above condition is much easier to realize than (5.25), in practice. The apparatus can also be modified in order to account for negative-result measurements. For these and other details, see Ref. 117.

The criteria (5.25) and (5.26) are special cases of Eq. (4.46). Note that here the decoherence parameter (for channel i) takes the value

$$\epsilon = 1 - \exp\left[-p_i^2(\delta D)^2/\hbar^2\right]. \tag{5.27}$$

The above model can also be considered as an explicit counterexample to the ergodic amplification theory,[45] discussed in Section 2.2 of Chapter 1, because the collapse of the wave function is obtained *without* the occurrence of any thermal irreversible process. Recall that the friction force has nothing to do with the occurrence of the wave function collapse, as mentioned at the beginning of this section.

2   One-dimensional emulsion model: The AgBr model

The "AgBr" model was originally introduced by Hepp in a remarkable paper[77] and is also known as the Coleman-Hepp model. It describes the interaction between an ultrarelativistic particle Q and a one-dimensional array D made up of N spin-1/2 objects. One can think, for instance, of a linear emulsion of AgBr molecules, where the down state of the spin corresponds to the undivided molecule and the up state to the dissociated molecule (Ag and Br atoms). The particle and each molecule interact via a spin-flipping local potential.

Let us start by giving the general Hamiltonian for this model. The total Hamiltonian H of the system composed of the above Q and D is

$$H = H_0 + H', \tag{5.28}$$

where

$$H_0 = H_Q + H_D \qquad (5.29)$$

is the free Hamiltonian and H' the interaction. Notice, in particular, that the free Hamiltonian of the ultrarelativistic particle Q is given by

$$H_Q = c\hat{p}, \qquad (5.30)$$

where $\hat{p}$ is the momentum operator of the particle: this choice of H_Q greatly simplifies the ensuing treatment and leads us to an exact expression for the S-matrix. Each spin-1/2 molecule making up the D system has two levels, which are usually taken to be the two eigenstates of the third Pauli matrix. In other words, H_D commutes with $\sigma_3^{(n)}$, where $\sigma_3^{(n)}$ is the third Pauli matrix acting on the nth ($n = 1, \ldots, N$) molecule. The interaction Hamiltonian H' is in general expressed as

$$H' = \int dx' \, \rho(\hat{x}, x'; \{\sigma\}) V(x') \qquad (5.31)$$

in terms of a local potential V and a density function ρ.[b] We denote by $\{\sigma\}$ the collection of spin operators (Pauli matrices) $\sigma^{(n)}$, each of which acts on the nth molecule and is responsible for the spin-flip interaction.

In order to calculate the S-matrix, we go to the interaction picture. In this picture, the above interaction Hamiltonian is transformed into

$$H_I'(t) = e^{iH_0 t/\hbar} H' e^{-iH_0 t/\hbar} = \int dx' \, \rho(\hat{x} + ct, x'; \{\sigma(t)\}) V(x'), \qquad (5.32)$$

where the explicit form of H_Q (5.30) has been used and

$$\sigma^{(n)}(t) = e^{iH_0 t/\hbar} \sigma^{(n)} e^{-iH_0 t/\hbar} = e^{iH_D t/\hbar} \sigma^{(n)} e^{-iH_D t/\hbar}. \qquad (5.33)$$

The evolution operator in the interaction picture $U_I(t) \equiv U_I(t, 0)$ satisfies [see (2.54)]

$$i\hbar \frac{d}{dt} U_I(t) = H_I'(t) U_I(t), \qquad (5.34)$$

which can be solved, under the initial condition $U_I(0) = 1$, to give

$$\begin{aligned} U_I(t) &= e^{-\frac{i}{\hbar} \int_0^t dt' H_I'(t')} \\ &= e^{-\frac{i}{\hbar} \int_0^t dt' \int dx' \, \rho(\hat{x} + ct', x'; \{\sigma(t')\}) V(x')}, \end{aligned} \qquad (5.35)$$

---

[b]The use of the symbol ρ for the *density function* will cause no confusion, even though we use the same symbol $\hat{\rho}$ to denote the *density matrix*.

as long as the commutability condition

$$[H_1'(t'), H_1'(t'')]$$
$$= \int dx' dx'' V(x') V(x'') [\rho(\hat{x} + ct', x'; \{\sigma(t')\}), \rho(\hat{x} + ct'', x''; \{\sigma(t'')\})]$$
$$= 0 \tag{5.36}$$

holds $\forall t'$ and t''. This condition is satisfied when the above commutator $[\rho, \rho]$ vanishes within the support of the potential V. Under the same condition, the S-matrix can be expressed as

$$S = e^{-\frac{i}{\hbar} \int_{-\infty}^{\infty} dt' \int dx' \, \rho(\hat{x} + ct', x'; \{\sigma(t')\}) V(x')}. \tag{5.37}$$

Suppose that we can choose a density function ρ that satisfies

$$\rho(\hat{x} + ct, x'; \{\sigma(t)\}) = \tilde{\rho}(\hat{x} + ct, x'; \{\sigma\}), \tag{5.38}$$

where $\tilde{\rho}$ is defined by this relation. Such a choice of ρ implies that the time dependence of $\sigma(t)$ is compensated by that of $\hat{x}(t) = \hat{x} + ct$ and that the possible t-dependence of ρ in the interaction picture appears only through the combination $\hat{x} + ct$. In this case, the above S-matrix can be further reduced to

$$S = e^{-\frac{i}{\hbar c} \int dx dx' \, \tilde{\rho}(x, x'; \{\sigma\}) V(x')}. \tag{5.39}$$

(Notice the absence of the Q operator $\hat{x}$ in the exponent.)

2.1 Introduction of the unitary-inequivalent representation

One of the simplest models for the one-dimensional emulsion system is undoubtedly described by the interaction Hamiltonian

$$H' = \sum_{n=1}^{N} V(\hat{x} - x_n) \sigma_1^{(n)} = \int dx' \rho(\hat{x}, x'; \{\sigma\}) V(x'), \tag{5.40}$$

$$\rho(\hat{x}, x'; \{\sigma\}) = \sum_{n=1}^{N} \delta(\hat{x} - x' - x_n) \sigma_1^{(n)}, \tag{5.41}$$

where $x_n = x_1 + (n - 1)d$ is the location of the nth AgBr molecule and N stands for the total number of molecules in the emulsion D. We have used $\sigma_1^{(n)}$ as a representative of an operator yielding spin-flipping: one can equally well use $\sigma_\pm^{(n)} = \frac{1}{2}(\sigma_1^{(n)} \pm i\sigma_2^{(n)})$. If, for simplicity, we neglect the free Hamiltonian H_D, then the conditions (5.36) and (5.38) are trivially satisfied and the above prescription immediately yields the S-matrix

$$S = \exp\left(-i\frac{\overline{V}\ell}{\hbar c} \Sigma_1^{(N)}\right), \tag{5.42}$$

where $\ell \equiv Nd$ and $\Sigma_i^{(N)} \equiv N^{-1} \sum_{n=1}^N \sigma_i^{(n)}$. Here we have also introduced the average potential strength $\overline{V}$ by

$$\overline{V}d \equiv \int dx V(x). \tag{5.43}$$

For simplicity and without a big loss of generality we have identified the "width" of the potential with the spacing d between adjacent molecules.

So far, N has been kept finite: It is interesting to see what happens in the infinite N limit, in connection with the MHS-like structure of D. In the infinite N limit, we have already seen (in Section 4 of Chapter 2) that the operators $\Sigma_i^{(N)}$ commute with each other and become c-numbers. In this case we regard ℓ as a continuous size parameter. In order to describe a macroscopic emulsion D, we have to introduce a distribution function $W(\ell)$ for ℓ, like in (4.41), centered at ℓ_0 with width $\Delta\ell$. According to the general prescription given in Chapter 4 [see, for example, (4.33) and (4.34) for the notation and the idea behind it] we face an integral of the type

$$\mathcal{W} \cdot e^{-i\kappa_0(\ell-\ell_0)} G = \int_{-\infty}^{\infty} d\ell\, e^{-i\kappa_0(\ell-\ell_0)} W(\ell)\langle\langle G(\ell)\rangle\rangle \tag{5.44}$$

for the interference term in a quantum measurement, which vanishes for

$$\kappa_0^2(\Delta\ell)^2 \gg 1. \tag{5.45}$$

The situation is quite similar to the case of the rigid mirror model. The vanishing of the above integral over ℓ yields the WFC, as we have stressed in Chapter 4.

Because $\sigma_1^{(n)}$ flips the nth spin, we can identify the c-number to which the average-spin operator $\Sigma_1^{(N)}$ tends in the infinite N limit, with the average transition probability of one molecule $q \equiv \sin^2(\overline{V}d/\hbar c)$ (see the discussion in the following subsections for details). In other words, we introduce the following transition

$$\Sigma_1^{(N)} \xrightarrow{N\to\infty} q \tag{5.46}$$

and exactly the same formula for $\Sigma_{\pm}^{(N)}$ apart from a trivial constant. This limiting procedure immediately yields

$$S \xrightarrow{N\to\infty} \overline{S} = e^{-i\kappa_0\ell} \tag{5.47}$$

with $\kappa_0 = (\overline{V}q/\hbar c)$. Equivalently, we have to introduce an MHS vacuum according to

$$\lim_{N\to\infty} \Sigma_1^{(N)}|\overline{0}\rangle = q|\overline{0}\rangle. \tag{5.48}$$

This is nothing but the transition to a unitary-inequivalent representation. We name $|\bar{0}\rangle$ the "MHS vacuum," after the Bogoliubov vacuum, to be described in the following.

In order to understand how this kind of transition in representation space is possible and to recognize its implication in physics, we first consider a bosonic case, in which the so-called Bogoliubov vacuum is a realization of a unitary-inequivalent representation. Let us introduce the Bogoliubov vacuum in which the annihilation and creation operators $\hat{a}_0$ and $\hat{a}_0^\dagger$ of a special mode 0 (corresponding to the minimum energy state) become c-number constants $\sqrt{N_0}$.[96,24] In other words, we define the Bogoliubov vacuum $|\bar{0}\rangle$ by

$$\hat{a}_0|\bar{0}\rangle = \hat{a}_0^\dagger|\bar{0}\rangle = \sqrt{N_0}|\bar{0}\rangle, \qquad \sqrt{N_0} = \langle\hat{a}_0\rangle_{\bar{0}} \equiv \langle\bar{0}|\hat{a}_0|\bar{0}\rangle \qquad (5.49)$$

with the same c-number N_0, the occupation number of mode 0. In the case of a finite medium, we have to assume that N_0 is constant inside the region occupied by the medium, but gradually and smoothly decreases to zero at the boundary. Of course, we also assume that the size of the medium is macroscopic, i.e. very large on a microscopic scale.

Observe that we cannot derive this type of representation of a field through a simple mathematical procedure, from the original number-diagonal representation, which is adherent to the commutation relation $[\hat{a}_0, \hat{a}_0^\dagger] = 1$. It is worth stressing that the introduction of the Bogoliubov vacuum is essentially motivated on physical, rather than mathematical grounds. We call this type of change of the representation *transition to an inequivalent representation*. When we go to a unitary-inequivalent representation, we usually expect a radically new physics, different from that in the original representation. It is important to notice that this type of transition is not unique, but a number of transitions are possible, and such a unitary-inequivalent transition, in other words, a transition from operators to c-numbers, is quite nontrivial mathematically. This procedure can be justified only on the basis of the physical results it predicts or on the kind of physical results one can derive from the relevant unitary-inequivalent representation: Only the entailing physics can justify the choice of the representation.

Applying the above technique to the original Hamiltonian, we easily derive the *reduced* interaction Hamiltonian, which is linear in the field operators.[96,24] The reduced interaction Hamiltonian yields the S-matrix

$$S = \exp\left[-i\sum_{k\neq 0}(\beta_k\hat{a}_k + \beta_k^*\hat{a}_k^\dagger)\right], \qquad (5.50)$$

with an appropriate c-number function β_k which is obtained from the original Hamiltonian. It is well known that the above S generates the famous Bogoliubov transformation. The reader can also refer to (2.109) of Chapter 2 in the case of coherent states. Of course, we can also introduce this kind of transformation in the one-dimensional emulsion model, by describing by means of an operator similar to (5.50) an excitation from the MHS vacuum; other details will not be discussed here.

We can also consider another example of unitary-inequivalent representation, i.e. the Higgs vacuum in field theory. As for the Higgs mechanism in the bosonic case, one usually introduces the Higgs vacuum (denoted by $|\overline{0}\rangle$) in which the operators $\hat{a}_0$ and $\hat{a}_0^\dagger$ of a special mode 0 become c-number constants over the whole space-time. In this case the c-number is related to the Higgs mass. Needless to say, the transition from the ordinary vacuum $|0\rangle$, defined by $\hat{a}_0|0\rangle = 0$, to the Higgs vacuum $|\overline{0}\rangle$ is a sort of transition to a unitary-inequivalent representation.

A more detailed treatment can be done in the case of our one-dimensional AgBr model, however it will be postponed to Section 2.3 since it requires more details of the model.

2.2   The Coleman-Hepp Hamiltonian and its modified version

From the above general Hamiltonian (5.28), the seminal Coleman-Hepp Hamiltonian for the Q + D system is obtained by neglecting the free Hamiltonian of the D system ($H_D = 0$) and by choosing the following density function

$$\rho(\hat{x}, x'; \{\sigma\}) = \sum_{n=1}^{N} \delta(\hat{x} - x' - x_n)\sigma_1^{(n)}, \tag{5.51}$$

which results in the total Hamiltonian

$$H = H_Q + H'^{(0)} = c\hat{p} + \sum_{n=1}^{N} V(\hat{x} - x_n)\sigma_1^{(n)}. \tag{5.52}$$

The above Hamiltonian was studied by several researchers,[18,138,118,109,135] because it nicely schematizes a quantum mechanical measurement process and can be solved exactly, as is easily understood from the vanishing commutation relation between the density functions and the validity of the relation (5.38) in this case. The model is indeed interesting and reflects several important features of a quantum measurement process. However, the interaction Hamiltonian in (5.52) is so simple that we can hardly deal with energy-transfer processes between D and Q. In order to improve this situation, we have to take into account the energy difference between the up and down spin states. One way of doing this[136] is to include the free Hamiltonian of the D system

$$H_D = \frac{1}{2}\hbar\omega \sum_{n=1}^{N} \left(1 + \sigma_3^{(n)}\right). \tag{5.53}$$

Moreover, it would also be interesting to discuss the energy-transfer problem by making use of a solvable model, described by a special interaction Hamiltonian. This is accomplished by choosing the following density function[136]

$$\rho(\hat{x}, x'; \{\sigma\}) = \sum_{n=1}^{N} \delta(\hat{x} - x' - x_n) \left(\sigma_+^{(n)} e^{-i\frac{\omega}{c}\hat{x}} + \sigma_-^{(n)} e^{i\frac{\omega}{c}\hat{x}} \right). \quad (5.54)$$

The above modifications make the model more realistic, even though the resulting Hamiltonian still remains a caricature of a linear "photographic emulsion" of AgBr molecules.

Thus the total Hamiltonian for the Q+D system reads

$$H = H_0 + H' = H_Q + H_D + H'$$

$$= c\hat{p} + \frac{1}{2}\hbar\omega \sum_{n=1}^{N} \left(1 + \sigma_3^{(n)} \right)$$

$$+ \sum_{n=1}^{N} V(\hat{x} - x_n) \left(\sigma_+^{(n)} e^{-i\frac{\omega}{c}\hat{x}} + \sigma_-^{(n)} e^{i\frac{\omega}{c}\hat{x}} \right), \quad (5.55)$$

where the notation is the same as in (5.52). Unlike the previous case, we are not neglecting the energy H_D of the array, namely the energy gap between the two states of each molecule. Notice that the energy difference between the two states of the molecule is $\hbar\omega$, and that the original model (5.52) is reobtained in the $\omega \to 0$ limit. The above Hamiltonian has attracted the attention of several researchers[186,78,114,112] due, in particular, to the presence of the free Hamiltonian of the array H_D, which enables it to take into account energy-exchange processes between Q and D: This is accomplished by the above interaction Hamiltonian, whose action can be decomposed in the following way

$$H'_{(n)}|p, \downarrow_{(n)}\rangle = V(\hat{x} - x_n)|p - \frac{\hbar\omega}{c}, \uparrow_{(n)}\rangle,$$

$$H'_{(n)}|p, \uparrow_{(n)}\rangle = V(\hat{x} - x_n)|p + \frac{\hbar\omega}{c}, \downarrow_{(n)}\rangle, \quad (5.56)$$

where $H'_{(n)}$ is the H'-term acting on the nth site, $|p, \downarrow_{(n)}\rangle$ represents a state in which the Q particle has momentum p and the nth molecule is undivided (spin down), and analogously for the other cases. We understand from Eq. (5.56) that the interaction Hamiltonian H' satisfies a "resonance condition," because the energy acquired or lost by the Q particle in every single interaction exactly matches the energy gap between the two spin states (i.e. the energy required to provoke one spin flip). Notice also that the choice of the density function in (5.54) has resulted in the fulfillment of conditions (5.38) and (5.36), keeping the model solvable.

We can develop the theoretical procedure in a way analogous to that described in the previous subsection and then obtain the MHS-theoretical description and explanation. For a while, however, we will analyze the present model by fixing N finite, in order to obtain a concrete image of the interaction and prepare the transition to the infinite N case.

The S-matrix is easily computed[136]

$$S^{[N]} = \exp\left(-i\frac{\overline{V}d}{\hbar c}\sum_{n=1}^{N}\sigma^{(n)}\cdot u\right) = \bigotimes_{n=1}^{N}\exp\left(-i\frac{\overline{V}d}{\hbar c}\sigma^{(n)}\cdot u\right), \qquad (5.57)$$

where $u \equiv [\cos(\omega x/c), \sin(\omega x/c), 0]$ and $\overline{V}d \equiv \int dx\, V(x)$, as in (5.43). This enables us to define the "spin-flip" probability, i.e. the probability of dissociating one AgBr molecule, as

$$q = \sin^2\left(\frac{\overline{V}d}{\hbar c}\right). \qquad (5.58)$$

The initial D state is taken to be the ground state $|0\rangle_N$ (N spins down), and we shall first consider the situation in which the initial Q state is a monochromatic plane wave $|p\rangle$. It is not difficult to compute the evolution

$$S^{[N]}|p,0\rangle_N = \sum_{j=0}^{N}\binom{N}{j}^{1/2}(-i\sqrt{q})^j\left(\sqrt{1-q}\right)^{N-j}|p-j\frac{\hbar\omega}{c},j\rangle_N, \qquad (5.59)$$

where $|p_j,j\rangle_N \equiv |p_j\rangle|j\rangle_N$ $(j = 1,\ldots,N)$, $|j\rangle_N$ being the symmetrized state in which j molecules are dissociated. As a consequence of the interaction, the Q particle loses j "quanta" of energy, by flipping j spins of the array.

In a typical interference experiment, a divider splits an incoming wave function ψ into two branch waves ψ_1 and ψ_2, so that the initial state of the Q+D system is

$$\Psi_I = (\psi_1 + \psi_2)|0\rangle_N, \qquad (5.60)$$

where $|\psi_i\rangle = \int dp_i\psi(p_i)|p_i\rangle$ $(i = 1, 2)$ are one-dimensional wave packets, normalized to unity. We assume that only ψ_2 interacts with D. The final state of the total system is

$$\Psi_F = |\psi_1\rangle|0\rangle_N + S^{[N]}|\psi_2\rangle|0\rangle_N \qquad (5.61)$$

and after recombination of the two branch waves the probability of observing the Q particle is

$$P = |\Psi_F|^2 = |\psi_1|^2 + |\psi_2|^2 + 2\text{Re}\left[{}_N\langle 0|\langle\psi_1|S^{[N]}|\psi_2\rangle|0\rangle_N\right]. \qquad (5.62)$$

Interference can be observed by inserting a phase shifter in one of the two paths (neutron-interferometer type), or when the two branch waves originating from the slits are forwarded to a distant screen (Young-interferometer type). In both cases, the visibility is readily calculated by (5.62)

$$V = \frac{P_{\text{MAX}} - P_{\text{min}}}{P_{\text{MAX}} + P_{\text{min}}} = \text{Re}\left[{}_N\langle 0|S^{[N]}|0\rangle_N \right] = \cos^N(\overline{V}d/\hbar c) = (1 - q)^{N/2},$$

$$(5.63)$$

where, for simplicity, we are suppressing the Q states. One can also compute several additional interesting quantities. For instance, the energy "stored" in the array after the interaction with the Q particle is

$$\langle H_{\text{D}}\rangle_{\text{F}} = {}_N\langle 0|S^{[N]\dagger} H_{\text{D}} S^{[N]}|0\rangle_N = qN\,\hbar\omega, \qquad (5.64)$$

where the subscript F stands for the final state (5.59). The fluctuation around the average is

$$\langle \Delta H_{\text{D}}\rangle_{\text{F}} = \sqrt{\langle (H_{\text{D}} - \langle H_{\text{D}}\rangle_{\text{F}})^2\rangle_{\text{F}}} = \sqrt{pqN}\,\hbar\omega, \qquad (5.65)$$

where $p = 1 - q$, and their ratio is given by

$$\frac{\langle \Delta H_{\text{D}}\rangle_{\text{F}}}{\langle H_{\text{D}}\rangle_{\text{F}}} = \sqrt{\frac{p}{qN}}. \qquad (5.66)$$

These are, of course, well-known properties of the binomial distribution. Although, in some sense, these results were already valid in the seminal Coleman-Hepp model (as far as the number of excited molecules is concerned), the quantities (5.64)-(5.66) could not be calculated (and would not have the same interesting physical meaning) in the original Hamiltonian (5.52), due to the absence of the free Hamiltonian H_{D}.

It is also possible to consider an initial thermal state for the spin system. Such a situation is more physical, in many respects, but the calculations are more involved.[136] Observe that the above results are exact and hold true for every value of N.

2.3 *Unitary-inequivalent representation in the AgBr model*

Let us scrutinize the possible appearance of a unitary-inequivalent representation in the $N \to \infty$ limit of the simplest AgBr model [Coleman-Hepp model, Eq. (5.52)], in accordance with the general procedure presented in Section 2.1. For simplicity, we first consider the emulsion as if it had a uniform distribution of AgBr molecules with a constant density. We have already seen (in Section 4 of Chapter 2) that the average-spin operators $\Sigma_{\pm}^{(N)}$ become c-numbers in the infinite N limit, which equivalently means that we can introduce an MHS vacuum $|\tilde{0}\rangle$: in this vacuum they are replaced with c-number constants whose values will be fixed only by the physics we are interested in. Remember that in the original Coleman-Hepp model, the S-matrix is expressed as [put $\omega = 0$ in (5.57)]

$$S^{[N]} = \exp\left(-i\frac{\overline{V}d}{\hbar c}\sum_{n=1}^{N}\sigma_1^{(n)}\right) = \exp\left(-iN\frac{\overline{V}d}{\hbar c}\Sigma_1^{(N)}\right)$$

$$= \bigotimes_{n=1}^{N}\exp\left(-i\frac{\overline{V}d}{\hbar c}\sigma_1^{(n)}\right) = \bigotimes_{n=1}^{N}\left[\sqrt{1-q} - i\sqrt{q}\,\sigma_1^{(n)}\right], \quad (5.67)$$

where q is the spin-flip probability defined in (5.58). If we prepare the initial D state in the ground state $|0\rangle_N$ (N spins down), the operator $(1 + \Sigma_3^{(N)})/2$ can be replaced with the spin-flip probability q, since it is expressed as the ratio of the number of up-spin states $\bar{n}$ over N

$$\frac{1}{2}\left(\Sigma_3^{(N)} + 1\right) = \frac{1}{N}\sum_{n=1}^{N}\frac{1}{2}\left(\sigma_3^{(n)} + 1\right) \longrightarrow \frac{\bar{n}}{N}, \quad (5.68)$$

which is nothing but q. From the commutation and anticommutation relations among $\Sigma_{\pm}^{(N)} = \frac{1}{2}\left(\Sigma_1^{(N)} \pm i\Sigma_2^{(N)}\right)$

$$[\Sigma_+^{(N)}, \Sigma_-^{(N)}] = \frac{1}{N}\Sigma_3^{(N)}, \qquad \{\Sigma_+^{(N)}, \Sigma_-^{(N)}\} = \frac{1}{N}, \quad (5.69)$$

we understand that the operators $\Sigma_{\pm}^{(N)}$ become commutable to each other in the $N \to \infty$ limit and therefore can be replaced with c-numbers $\sqrt{q/N}$ because of the relation

$$\Sigma_+^{(N)}\Sigma_-^{(N)} = \frac{1}{N}\frac{1}{2}\left(\Sigma_3^{(N)} + 1\right) \longrightarrow \frac{q}{N}. \quad (5.70)$$

In other words, we have introduced an MHS vacuum $|\overline{0}\rangle$, defined by the relations

$$\Sigma_{\pm}^{(N)}|\overline{0}\rangle = \sqrt{q/N}e^{\pm i\varphi}|\overline{0}\rangle, \qquad \sqrt{q/N}e^{\pm i\varphi} \equiv \langle\overline{0}|\Sigma_{\pm}^{(N)}|\overline{0}\rangle, \quad (5.71)$$

over the space-time region occupied by the array, where φ stands for a possible phase. In this way, we can deal with the transition to a unitary-inequivalent representation, by which we get

$$\Sigma_{\pm}^{(N)} \longrightarrow \sqrt{q/N}e^{\pm i\varphi}, \quad (5.72)$$

and therefore a *c-number S-matrix*

$$S^{[N]} = \exp\left(-iN\frac{\overline{V}d}{\hbar c}\Sigma_1^{(N)}\right)$$

$$\longrightarrow \exp\left(-i\gamma N\frac{\overline{V}d}{\hbar c}\sqrt{\frac{q}{N}}\right) = \exp\left(-i\gamma\sqrt{N}\frac{\overline{V}d}{\hbar c}\sin\frac{\overline{V}d}{\hbar c}\right), \quad (5.73)$$

as N goes to infinity, where γ is a (real) constant. It is important to notice that this expression is meaningful only when the exponent has a well-defined limit. That is, the choice of the above MHS vacuum $|\overline{0}\rangle$ in (5.71) necessarily requires the potential $V(x)$ to behave like ($\int dx\, V(x) \equiv \overline{V}d$)

$$\frac{\overline{V}d}{\hbar c} \sin \frac{\overline{V}d}{\hbar c} \sim O(1/\sqrt{N}), \tag{5.74}$$

in the $N \to \infty$ limit, otherwise we would end up with a trivial ($S = 1$) or an ill-defined S-matrix: The introduction of $|\overline{0}\rangle$ in (5.71) is justifiable provided we are interested in the physics prescribed by the above condition (5.74). Observe also that in the weak-potential limit, Eq. (5.74) implies the behaviours

$$\frac{\overline{V}d}{\hbar c} \longrightarrow O(1/\sqrt[4]{N}), \quad q \equiv \sin^2 \frac{\overline{V}d}{\hbar c} \longrightarrow O(1/\sqrt{N}), \quad \bar{n} = qN \longrightarrow O(\sqrt{N}),$$
$$\tag{5.75}$$

which are to be contrasted with those in the so-called "weak-coupling, macroscopic limit" to be introduced in the subsequent section.

It is interesting to see that the above expression of the S-matrix (5.73) can be taken as a starting point for the MHS treatment of our one-dimensional emulsion model. Notice that the emulsion length is written as $\ell = Nd$ and we can rewrite the S-matrix in (5.73) as

$$S^{[N]} \longrightarrow e^{-i\gamma\sqrt{q/N}(\overline{V}/c)\ell/\hbar} \equiv e^{-i\kappa_0\ell}, \tag{5.76}$$

which is just the same form as in (5.47) and is nothing but the general form of the S-matrix we have encountered in Chapter 4. For the reason mentioned in Section 2.1, ℓ is distributed with the Gaussian distribution function $W(\ell)$, centered at ℓ_0 with width $\Delta\ell$.

We have to keep in mind that what has been shown above is just one example of realization of a unitary-inequivalent representation: There are actually many realizations, depending on how the limiting procedure $N \to \infty$ is taken. A different example (the so-called *weak-coupling, macroscopic limit*) will be seen in the following subsections. We stress again that the appearance of an inequivalent representation is an important characteristic of systems with infinite degrees of freedom.

2.4   The weak-coupling, macroscopic limit

It is interesting to point out some properties of the spin system in the AgBr model. As we have seen, the average "spin" of the array

$$\Sigma_i^{(N)} = \frac{1}{N} \sum_{n=1}^{N} \sigma_i^{(n)}, \quad i = 1, 2, 3 \tag{5.77}$$

behaves as a classical quantity in the infinite N limit. Indeed, the commutation relation

$$\left[\Sigma_i^{(N)}, \Sigma_j^{(N)}\right] = 2i\frac{1}{N}\Sigma_k^{(N)} \tag{5.78}$$

(with i, j, k cyclic permutation of $1, 2, 3$) vanishes in the $N \to \infty$ limit, so that these operators tend to the center of the von Neumann algebra of observables of the array (see Section 4 of Chapter 2).

Notice that the total spin

$$N\Sigma_j^{(N)} = \sum_{n=1}^{N} \sigma_j^{(n)}, \qquad j = 1, 2, 3 \tag{5.79}$$

is always a fully quantum mechanical object (as it should be), because

$$\left[N\Sigma_i^{(N)}, N\Sigma_j^{(N)}\right] = 2iN\Sigma_k^{(N)} \tag{5.80}$$

(i, j, k even permutations of $1, 2, 3$).

However, there is another remarkable situation. Consider the weak-coupling ($q \to 0$), macroscopic ($N \to \infty$) limit, with $qN \equiv \bar{n} =$ finite.[136] The physical meaning of this limit is appealing: It corresponds to admitting that the number of dissociated molecules $\bar{n} = qN$ is finite. Alternatively, one can say that the energy $\bar{n}\hbar\omega = qN\hbar\omega$ exchanged between the particle and the detector is kept finite, even though the number of elementary constituents of D becomes very large. As we shall see, such a limit is closely related to Van Hove's "$\lambda^2 T$" limit.[191]

Let us evaluate the physical quantities calculated in the previous subsection in the $q \to 0$, $N \to \infty$, $\bar{n} = qN =$ finite limit. From (5.63)-(5.66) one obtains

$$\begin{aligned}
\langle H_D \rangle_F &\to \bar{n}\,\hbar\omega, \\
\langle \Delta H_D \rangle_F &\to \sqrt{\bar{n}}\,\hbar\omega, \\
\frac{\langle \Delta H_D \rangle_F}{\langle H_D \rangle_F} &\to \frac{1}{\sqrt{\bar{n}}}, \\
\mathcal{V} &\to e^{-\bar{n}/2}.
\end{aligned} \tag{5.81}$$

This is nothing but the Poisson limit of the binomial distribution displayed in Section 2.2.

The simple macroscopic limit $N \to \infty$, with $q \neq 0$, would yield only divergent or vanishing quantities. But there is more: Perform the following change of basis for the generators of $SU(2)$

$$\begin{pmatrix} h_+ \\ h_- \\ h_3 \\ 1 \end{pmatrix} \equiv \begin{pmatrix} N^{-1/2} & & & \\ & N^{-1/2} & & \\ & & 1 & N/2 \\ & & & 1 \end{pmatrix} \begin{pmatrix} N\Sigma_+^{(N)} \\ N\Sigma_-^{(N)} \\ N\Sigma_3^{(N)}/2 \\ 1^{(N)} \end{pmatrix}. \tag{5.82}$$

The commutation properties for h, 1 are

$$[h_3, h_\pm] = \pm h_\pm,$$

$$[h_-, h_+] = 1 - \frac{2}{N} h_3,$$

$$[h, 1] = 0, \tag{5.83}$$

and yield, in the $N \to \infty$ limit, the standard boson commutation relations. Thus the following identification follows:

$$h_+ = \frac{1}{\sqrt{N}} \sum_{n=1}^{N} \sigma_+^{(n)} = \sqrt{N} \Sigma_+^{(N)} \overset{N \to \infty}{\longrightarrow} a^\dagger,$$

$$h_- = \frac{1}{\sqrt{N}} \sum_{n=1}^{N} \sigma_-^{(n)} = \sqrt{N} \Sigma_-^{(N)} \overset{N \to \infty}{\longrightarrow} a, \tag{5.84}$$

$$h_3 = \frac{1}{2} \sum_{n=1}^{N} \left(1 + \sigma_3^{(n)}\right) = \frac{N}{2} \left(1^{(N)} + \Sigma_3^{(N)}\right) \overset{N \to \infty}{\longrightarrow} \mathcal{N} \equiv a^\dagger a.$$

The weak-coupling, macroscopic limit of the scattering matrix $S^{[N]}$ in (5.57) is also remarkable. A short manipulation yields

$$S^{[N]} \to S = \exp\left[-i\frac{u_0 d}{\hbar c}\left(a^\dagger \exp\left[-i\frac{\omega}{c}\hat{x}\right] + a \exp\left[i\frac{\omega}{c}\hat{x}\right]\right)\right], \tag{5.85}$$

where $u_0 \equiv V\sqrt{N}$ and therefore $u_0 d/\hbar c \propto \sqrt{qN} = \sqrt{\bar{n}} < \infty$. The connection with a "maser" is manifest, and has been elucidated in Ref. 136. This is to be compared with (5.50) and is a typical result of the transition to a unitary-inequivalent representation.

Equations (5.81), (5.84) and (5.85) can again be regarded as a sort of transition to the unitary-inequivalent representation, within the MHS-theoretical framework. In such a case, we can follow the procedure outlined in the first part of this section. The above limiting procedure, however, does *not* lead the operators involved to the center of the algebra of observables of the D system. In this sense, one might say that the $N \to \infty$, $qN = \bar{n} =$ finite limit "foreruns" the appearance of a superselection-rule space. In order to properly address this issue, we need to investigate in more detail the interaction between Q and D. In particular, the study of the temporal behaviour of the total (Q+D) system turns out to be very interesting: We shall see in the next subsection that this analysis will bring to light the presence of a Wiener process.

2.5 *Temporal behaviour of a propagator*

A quantum measurement is essentially an irreversible process. On the other hand, it is well known that a dissipative dynamics should be characterized by exponential laws, which are representative of irreversible phenomena. (Notice that the quantum mechanical situation is far from being trivial, as discussed in the Appendix.) It is therefore worth investigating the temporal behaviour of our AgBr system, in order to understand whether it displays genuine irreversible features.

The temporal evolution of the AgBr model is best disclosed by studying the behaviour of the propagator. We will only present here the main results and discuss their physical meaning. A complete analysis is given in Refs. 136, 23.

Observe first that the interaction Hamiltonian H' has nonvanishing matrix elements only between those eigenstates of H_0 whose spin-quantum numbers differ by one. Furthermore, since (5.56) describes an energy-exchange process, we can expect a dissipation effect of some sort.

It is not difficult to show that the evolution operator is disentangled like

$$
e^{-iHt/\hbar} = e^{-iH_0t/\hbar} \prod_{n=1}^{N} \left(e^{-i\tan(\alpha_n)\sigma_+^{(n)}(\hat{x})} e^{-\ln\cos(\alpha_n)\sigma_3^{(n)}} e^{-i\tan(\alpha_n)\sigma_-^{(n)}(\hat{x})} \right),
$$

(5.86)

where

$$
\alpha_n \equiv \alpha_n(\hat{x}, t) \equiv \int_0^t V(\hat{x} + ct' - x_n)dt'/\hbar, \tag{5.87}
$$

$$
\sigma_\pm^{(n)}(\hat{x}) \equiv \sigma_\pm^{(n)} \exp\left(\mp i\frac{\omega}{c}\hat{x}\right). \tag{5.88}
$$

We shall concentrate our attention on the situation in which the Q particle is initially at the position $x' < x_1$, where x_1 is the position of the first scatterer in the linear array, and is moving towards the array with speed c. As before, the spin system is initially set in the ground state $|0\rangle_N$ of the free Hamiltonian H_D (all spins down). This choice of the ground state is meaningful, from a physical point of view, because the Q particle is initially outside D.

The propagator

$$
G(x, x', t) \equiv \langle x| \otimes {}_N\langle 0|e^{-iHt/\hbar}|0\rangle_N \otimes |x'\rangle, \tag{5.89}
$$

can be easily calculated from (5.86) to be

$$
G(x, x', t) = \langle x| \otimes {}_N\langle 0|e^{-ic\hat{p}t/\hbar} \prod_{n=1}^{N} \left(e^{-\ln\{\cos[\alpha_n(\hat{x}, t)]\}\sigma_3^{(n)}} \right) |0\rangle_N \otimes |x'\rangle
$$

$$
= \langle x|x' + ct\rangle \prod_{n=1}^{N} \left(e^{\ln\{\cos[\alpha_n(x', t)]\}} \right)
$$

$$
= \delta(x - x' - ct) \prod_{n=1}^{N} \cos\alpha_n(x', t). \tag{5.90}
$$

Observe that, due to the choice of the free Hamiltonian H_Q in Eq. (5.55), the Q wave packet does not disperse and moves with constant speed c. We place the spin array at the far right of the origin ($x_1 \gg 0$) and consider the case where potential V has a compact support and the Q particle is initially located at the origin $x' = 0$, i.e. well outside the potential region of D. The above equation shows that the evolution of Q occurs only along the path $x = ct$. Therefore we obtain

$$G(x, 0, t) = \delta(x - ct) \prod_{n=1}^{N} \cos \tilde{\alpha}_n(t), \qquad \tilde{\alpha}_n(t) \equiv \int_0^{ct} V(y - x_n) dy / \hbar c. \quad (5.91)$$

This result is exact. Notice that the spin-flip probability (5.58) can be written

$$q \equiv \sin^2 \tilde{\alpha}_n(\infty) = \sin^2 \left(\frac{\overline{V} d}{\hbar c} \right). \quad (5.92)$$

We consider again the weak-coupling, macroscopic limit

$$q \simeq \left(\frac{\overline{V} d}{\hbar c} \right)^2 = O(N^{-1}). \quad (5.93)$$

Observe that if we set

$$x_n = x_1 + (n - 1)d, \qquad L = x_N - x_1 = (N - 1)d, \quad (5.94)$$

where L stands for the total size of the emulsion (and is not to be confused with $L = P'/f$ introduced at the beginning of the preceding section), a summation over n is to be replaced by a definite integration

$$q \sum_{n=1}^{N} f(x_n) \to \frac{q}{d} \int_{x_1}^{x_N} f(z) dz \simeq \frac{\overline{n}}{L} \int_{x_1}^{x_N} f(z) dz. \quad (5.95)$$

This integration gives a finite result if the function f is scale invariant, because the integration volume is considered to be finite from the physical point of view: The scaled variable $z_n \equiv x_n / L$ can be considered as a continuous variable z in the above limit, for $d/L \to 0$ as $N \to \infty$. Indeed, the quantities x_1/L and x_N/L should be of the order of unity even in the $L \to \infty$ limit. It will be shown below [Eq. (5.96)] that in the present case the function f is indeed scale invariant.

For the sake of simplicity, we shall work in the Fermi-Yang approximation $V(y) = (\overline{V} d)\delta(y)$. We obtain

$$G \propto \exp \left(\sum_{n=1}^{N} \ln \left\{ \cos \int_{-x_n}^{ct - x_n} (\overline{V} d / \hbar c) \delta(y) dy \right\} \right)$$

$$= \exp\left(\sum_{n=1}^{N} \ln\left\{\cos\left[(\overline{V}d/\hbar c)\theta(ct - x_n)\right]\right\}\right)$$

$$\rightarrow \exp\left(-\frac{\overline{n}}{2}\int_{x_1}^{x_n}\theta(ct - z)dz\right)$$

$$= \exp\left(-\frac{\overline{n}}{2}\left[\frac{ct - x_1}{L}\theta(x_N - ct)\theta(ct - x_1) + \theta(ct - x_N)\right]\right), \quad (5.96)$$

where θ is the step function and the arrow denotes the weak-coupling, macroscopic limit (5.93)-(5.95). This brings about an exponential regime as soon as the interaction starts. Indeed, if $x_1 < ct < x_N$,

$$G \propto \exp\left(-\overline{n}\frac{c(t - t_0)}{2L}\right), \quad (5.97)$$

where $t_0 = x_1/c$ is the time at which the Q particle meets the first potential. Notice that there is *no* Gaussian behaviour at short times and *no* power law at long times (see the Appendix).

It is well known[132] that deviations from the exponential behaviour at short times are a consequence of the finiteness of the mean energy of the initial state. If the position eigenstates in Eq. (5.89) are substituted with wave packets of size a, a detailed calculation shows that the exponential regime is attained a short time after t_0, of the order of a/c, which, in the present model, can be made arbitrarily small. The region $t \sim t_0 + O(a/c)$ may be viewed as a possible residuum of the short-time Gaussian-like behaviour. For this reason, the temporal behaviour just derived is not in contradiction with well-known general theorems.[99,58]

It is interesting to bring to light the profound link between the weak-coupling, macroscopic limit $qN = \overline{n} =$ finite considered in this subsection and Van Hove's "$\lambda^2 T$" limit.[191] Van Hove's so-called diagonal singularity naturally appears in the present model. It is easy to check that for each diagonal matrix element of H'^2 there are N intermediate-state contributions: Indeed, for example

$$\langle 0, \ldots, 0|H'^2|0, \ldots, 0\rangle = \sum_{j=1}^{N}|\langle 0, \ldots, 0|H'|0, \ldots, 0, 1_{(j)}, 0, \ldots, 0\rangle|^2. \quad (5.98)$$

On the other hand, at most 2 states can contribute to each off-diagonal matrix element of H'^2. This ensures that only the diagonal matrix elements are kept in the weak-coupling, macroscopic limit with $qN < \infty$, which is the realization of diagonal singularity in our model. The link with the $\lambda^2 T$ limit is easily evinced by observing that q, in (5.93), is nothing but the square of a coupling constant (Van Hove's λ), and that $N(\propto L)$ can be considered proportional to the total interaction time T. Notice that the "lattice spacing" d, the inverse of which corresponds to a density in our one-dimensional model, can be kept finite in the limit. (In such a case, we have to express everything in terms of scaled variables, that is, $\tau \equiv t/L$, z_1 and z_N, introduced after Eq. (5.95) and $\zeta \equiv a/L$, where a is the size of the wave packet.) One can therefore expect the presence of a dissipative effect of some sort.

The occurrence of a stochastic process in the modified AgBr system was brought to light in Ref. 23. Let us only give the main results. The initial condition will be taken just the same as before: We place the Q particle well outside D, moving towards D with speed c, while the spin system is set in the ground state $|0\rangle_N$ of the free Hamiltonian H_D (all spins down). The initial state is written

$$|\psi, 0\rangle_N \equiv |\psi\rangle \otimes |0\rangle_N = \int dx \psi(x)|x\rangle \otimes |0\rangle_N, \qquad \int_{-\infty}^{+\infty} dx |\psi(x)|^2 = 1, \quad (5.99)$$

where, for simplicity, we choose ψ to be symmetrically distributed around the origin and with a compact support. A possible (and convenient) choice is $\psi(x) = (2a)^{-1/2}\theta(a - |x|)e^{ip_0 x/\hbar}$, and we require

$$a \ll x_1, \qquad a \ll L \equiv x_N - x_1. \tag{5.100}$$

We consider again the weak-coupling, macroscopic limit (5.93), in the ansatz (5.94)-(5.95), and the situation in which Q is *inside* D, and work in the Fermi-Yang approximation $V(y) = (\overline{V}d)\delta(y)$. A straightforward, if lengthy, calculation yields

$$\langle c\hat{p}(t)\rangle \equiv {}_N\langle 0, \psi|c\hat{p}(t)|\psi, 0\rangle_N$$
$$= cp_0 - \hbar\omega \frac{\bar{n}}{L}(ct - x_1) + \text{b.e.}, \tag{5.101}$$

where $\hat{p}(t)$ is the momentum of Q in the Heisenberg picture and b.e. is a short-hand notation for "border effects," namely terms appearing for $|ct - x_1|$, $|ct - x_N| \leq a$. Equation (5.101) displays an energy dissipative process.

Consider now the operator

$$\Delta H_D(t) \equiv H_D(t) - H_D(0). \tag{5.102}$$

Its expectation value in the state (5.99) is

$$\langle \Delta H_D(t)\rangle \equiv {}_N\langle 0, \psi|\Delta H_D(t)|\psi, 0\rangle_N = \hbar\omega \frac{\bar{n}}{L}(ct - x_1) + \text{b.e.}, \tag{5.103}$$

in agreement with (5.101) (the energy lost by Q must be stored in D). Finally, define the operator

$$\Sigma(t) = \Delta H_D(t) - \langle \Delta H_D(t)\rangle. \tag{5.104}$$

One obtains, after some manipulation,

$$\langle \Sigma(t) \rangle = 0, \tag{5.105}$$

$$\langle \Sigma(t_1)\Sigma(t_2) \rangle = (\hbar\omega)^2 \frac{\bar{n}}{L}[c\min(t_1, t_2) - x_1] + \text{b.e.}$$

$$= (\hbar\omega)^2 \frac{\bar{n}}{L} c\min(\tau_1, \tau_2) + \text{b.e.}, \tag{5.106}$$

where $\tau_j = t_j - x_1/c$ $(j = 1, 2)$ and we neglected the size of the wave packet as compared to the size of D [see (5.100)].

All these results are valid in the weak-coupling, macroscopic limit, in the restricted Hilbert space spanned by $|\psi, 0\rangle_N$. The properties (5.105) and (5.106) are reminiscent of a Wiener process. However, in order to shed light on the presence and the nature of the stochastic process, one must show that the process is Gaussian, by computing the multi-time correlation functions. This can be accomplished by defining the characteristic functional

$$\phi[\beta] \equiv \langle e^{\int dt\beta(t)\Sigma(t)} \rangle \equiv {}_N\langle 0, \psi | e^{\int dt\beta(t)\Sigma(t)} |\psi, 0\rangle_N. \tag{5.107}$$

The disentanglement of the exponential is straightforward but somewhat involved, and requires some care. The final result is[23]

$$\phi[\beta] \longrightarrow e^{\frac{1}{2}(\hbar\omega)^2 \frac{\bar{n}}{L} \int dt dt'\beta(t)[c\min(t,t')-x_1]\beta(t')}$$

$$= e^{\frac{1}{2}(\hbar\omega)^2 \frac{\bar{n}}{L} \int dt dt'\beta(t)c\min(\tau,\tau')\beta(t')}, \quad (\tau \equiv t - x_1/c). \tag{5.108}$$

The characteristic functional turns out to be Gaussian, which proves that $\Sigma(t)$ behaves as a Gaussian stochastic process, and that, due to (5.106), it is nothing but a Wiener process, in the weak-coupling, macroscopic limit (5.93)-(5.95), within the restricted state space spanned by $|\psi, 0\rangle_N$.

The operator $\Sigma(t)$, which is essentially the energy operator $H_D(t)$ for the N-spin array D, serves as a "noise operator" in the $N \to \infty$ limit (5.93), in the sense of Eqs. (5.105) and (5.106). Furthermore, there explicitly appears a Wiener stochastic process, characterized by the Gaussian white noise properties or by the above characteristic functional (5.108). Although the appearance of a stochastic process of some sort is not entirely surprising, and could somehow be expected on the basis of the behaviour of the propagator (5.97), the emergence of the Gaussian white noise is remarkable, for such a nontrivial Hamiltonian like (5.55). We understand that the weak-coupling, macroscopic limit (5.93)-(5.95), which is closely related to Van Hove's limit, plays a crucial role in this respect: Physically, it corresponds to a kind of coarse graining and scale-changing procedure. Another important fact, not to be dismissed, is that we have exclusively considered the dynamics within a restricted state space spanned by the initial state $|\psi, 0\rangle_N$ in (5.99). In relation to these points, one has to pay special attention to the limiting process $N \to \infty$, since such a macroscopic limit is expected to make the representation space unitarily inequivalent to the original one with finite N. This would be the place where one can think of a possible relevance of the AgBr model to the MHS theory.

Notice also that the role played by probabilistic considerations is completely different, in classical and quantum mechanics. In classical mechanics, probabilities are epistemic and essentially arise from our ignorance on the initial conditions. By contrast, quantum mechanical probabilities are ontological and nonepistemic, and constitute a fundamental part of the very structure of the theory. This makes the derivation of a dissipative dynamics in quantum mechanics very different from its classical counterpart. In particular, we feel that a true understanding of a quantum dissipative behaviour will eventually deepen our insight into the quantum measurement problem and its intrinsic irreversibility. At the same time, the solution of the above-mentioned issues might enable us to put forward a proper definition of "probability dissipation" in quantum mechanics, along the line of thought of Ref. 130. Clearly, these are only speculations, that have not yet been corroborated by clear-cut arguments, at the present stage of investigation.

The model discussed in this section has proven to be a very fertile example for discussions on quantum measurements and dissipation. Even though many problems remain open, the model explicitly shows that it is possible to view a quantum mechanical measurement as a sort of irreversible and dissipative process.

3   Cini's model

Let us now focus our attention on another model Hamiltonian, originally put forward by Cini.[37] In this approach, the evolution from a pure to a mixed state is ascribed to the second law of thermodynamics, but unlike in the ergodic amplification theory,[45] discussed in Section 2.2 of Chapter 1, no problem arises in connection with negative-result measurements, which can be tackled in a consistent way. It was shown[135] that Cini's approach can be "blended" to the MHS theory, by introducing statistical fluctuations into his model. We will first briefly summarize the most salient points of Cini's discussion, and will then show the modifications when the MHS structure of the detector is taken into account.

Consider a two-level quantum system Q, described by the state

$$\chi_Q = (c_+ u_+ + c_- u_-)\phi(\mathbf{r}), \qquad (5.109)$$

where $|c_+|^2 + |c_-|^2 = 1$, $u_\pm$ are the eigenstates of τ_3, the third Pauli matrix, and $\phi(\mathbf{r})$ is the position wave packet. After the spectral decomposition, the system is described by

$$\chi_Q = c_+ u_+ \phi_+(r) + c_- u_- \phi_-(r), \tag{5.110}$$

where $\phi_\pm(r)$ are distinct wave packets, travelling in different regions of space. The quantum system interacts with a detector D, made up of N particles, which are also assumed to have only two possible states, say a ground state ω_0 and an excited or ionized state ω_1. The interaction Hamiltonian is assumed to be

$$H_1 = \frac{g_0}{\sqrt{N}} \frac{1}{2}(1 + \tau_3)(a_0^\dagger a_1 + a_1^\dagger a_0), \tag{5.111}$$

where g_0 is a coupling constant and $a_0^\dagger$ and $a_1^\dagger$ are the creation operators for the states ω_0, ω_1, respectively, and obey boson commutation relations: $[a_0, a_0^\dagger] = [a_1, a_1^\dagger] = 1$. All other commutators vanish. Due to the term $\frac{1}{2}(1 + \tau_3)$, only the u_+ component of χ_Q yields a nonvanishing interaction. A state of D is defined by assigning the number n of particles in the ground state ω_0:

$$|n, N - n\rangle = \frac{1}{\sqrt{n!}} \frac{1}{\sqrt{(N-n)!}} (a_0^\dagger)^n (a_1^\dagger)^{N-n} |0\rangle, \tag{5.112}$$

where $N - n$ is the number of particles in the excited state ω_1.

The total system Q+D is initially described by the vector

$$\psi(0) = \chi_Q \otimes |N, 0\rangle, \tag{5.113}$$

and its evolution is given by

$$\psi(t) = e^{-\frac{i}{\hbar}H_1 t}\psi(0) = c_+ u_+ \phi_+(r) \sum_{n=0}^{N} a_n(t)|n, N - n\rangle + c_- u_- \phi_-(r)|N, 0\rangle,$$

$$a_n(t) = \frac{\sqrt{N!}(-i)^{N-n}}{\sqrt{n!}\sqrt{(N-n)!}} \cos^n \frac{gt}{\hbar} \sin^{N-n} \frac{gt}{\hbar}, \tag{5.114}$$

where $g = g_0/\sqrt{N}$. The probability of finding n particles in the ground state at time t is therefore given by

$$P_n(t) = |a_n(t)|^2 = \binom{N}{n} p(t)^n q(t)^{N-n}, \tag{5.115}$$

where $p(t) = \cos^2 \alpha(t)$, $q(t) = 1 - p(t) = \sin^2 \alpha(t)$, $\alpha(t) = gt/\hbar$. For N large, the distribution of P_n is very strongly peaked around its maximum

$$\bar{n}(t) = Np(t), \tag{5.116}$$

and application of Stirling's formula to Eq. (5.115) gives

$$P_{\bar{n}} \simeq 1. \tag{5.117}$$

Since, however, $\sum_{n=0}^{N} P_n = 1$, the probability of finding $n \neq \bar{n}(t)$ is negligible, and one can approximate Eq. (5.114) with

$$\psi(t) \simeq c_+ u_+ |\bar{n}(t), N - \bar{n}(t)\rangle + c_- u_- |N, 0\rangle. \tag{5.118}$$

Note that, from Eqs. (5.116) and (5.117), one gets $P_n(0) = \delta_{nN}$, $P_n(t_0) = \delta_{n0}$, with

$$t_0 = \pi\hbar\sqrt{N}/2g_0, \tag{5.119}$$

so that t_0, the time for complete discharge of D, is proportional to $N^{1/2}$. This is the reason why the coupling constant, in Eq. (5.111), contained a $N^{-1/2}$ factor. More precisely, Eq. (5.115) yields

$$P_{\bar{n}+\Delta n} = \frac{1}{\sqrt{2\pi N pq}} e^{-(\Delta n)^2/2Npq}, \tag{5.120}$$

for the probability of finding $\bar{n} + \Delta n$ particles in the ground state, at any given time t.

Cini's interesting analysis goes much further, in terms of the density matrix of the total system Q+D. However, we will limit ourselves to his discussion on the case of an *irreversible* detector. Obviously, for $t = 2t_0$, by Eq. (5.116), one obtains $P_n(2t_0) = \delta_{nN}$, so that the system eventually goes back to its initial state. This is due to the hermiticity of the Hamiltonian H_1, that describes a *reversible* interaction. Cini solves this problem by invoking the second law of thermodynamics: According to him, "no counter will ever recharge itself because [at $t = t_0$] all the electrons and ions will spontaneously recombine to form the initial neutral state."

In the following, we will give this statement a precise mathematical expression, by following an alternative approach. Moreover, we will argue that the decoherence of the Q+D system is ascribable to the MHS structure of D.

In the above analysis, the spatial degrees of freedom are completely neglected. The quantum system Q and all the particles of the detector D occupy the same point of space, for a certain "interaction" time t_0, defined in Eq. (5.119), during which the interaction described by H_1 takes place. We shall now show what happens when space is taken into account. Let us start from the Hamiltonian

$$H = H_Q + H_D + H_2, \tag{5.121}$$

where H_Q is the free Hamiltonian of the system Q, H_D the free Hamiltonian of the detector D, and H_2 the interaction Hamiltonian. A possible choice is:

$$H_Q = p^2/2m, \quad H_D = \frac{1}{2}\hbar\omega \sum_{n=1}^{N}(1 + \sigma_3^{(n)}), \quad H_2 = \frac{1}{2}(1 + \tau_3) \sum_{n=1}^{N} V(r - r_n)\sigma_1^{(n)},$$

(5.122)

which describes a particle at position r, with momentum p, interacting with a detector made up of N (AgBr) molecules. The molecules' positions are r_n, $(n = 1, ..., N)$, $\sigma_1^{(n)}$ and $\sigma_3^{(n)}$ are Pauli matrices acting on the n-th molecule, and ω is a constant. As already stressed after (5.111), due to the term $\frac{1}{2}(1+\tau_3)$ only the u_+ component of χ_Q, in (5.109), yields a nonvanishing interaction. The Hamiltonian H_2 is essentially the same as the one introduced in (5.52), the only difference being that we are here considering the three-dimensional case, in which $r, r_n \in \mathbf{R}^3$.

It can be shown[135] that the Hamiltonian H in (5.121) can be identified with H_1 in (5.111) in the limit $\omega = 0$ and the particle Q is at rest in the detector: This is accomplished by making the hypothesis of "broad" potentials $V(r - r_n) = V_0\delta\Omega\rho(r)/N$, where $\rho(r) = \sum_{n=1}^{N} \delta^3(r - r_n)$ is the density of particles per unit volume, with $\int d^3r\rho(r) = N$. In this way, the detector D is transformed into a "fluid" of density $\rho(r)$, in which up to N excitations can be created.[c] If now we were to substitute $\rho(r)$ with a constant, "average" value, the original Hamiltonian H would lose any dependence on the molecules' locations. In doing so, we would overestimate the quantum-mechanical coherence of the detector, because we would neglect the additional phase randomization actually introduced by the neglected spatial degrees of freedom. In order to better take into account the random motion of the elementary constituents of D, we substitute the operator ρ with a c-number plus a fluctuation

$$\rho(r) \rightarrow \langle \rho \rangle + \delta\rho(r),$$

(5.123)

where $\langle \rho \rangle$ is a constant, "background" density, and $\delta\rho$ obeys the statistical properties

$$\langle \delta\rho(r) \rangle = 0, \quad \langle (\delta\rho(r))^2 \rangle = F_\theta,$$

(5.124)

where the brackets denote an ensemble average over all the possible microscopic configurations of the detector, and θ is the detector temperature. The precise mathematical expression for F_θ depends on the physical and chemical properties of D. For our analysis it will suffice to know that the behaviour of F_θ becomes critical when a phase transition takes place. We wish to stress that it is precisely at this point that the MHS structure of D comes into play: Since it is a macroscopic system, D cannot be described within a single Hilbert space,

---

[c]The identification of the interaction Hamiltonian H_2, in (5.122) and H_1, in (5.111), has been sharpened in Ref. 137.

and its MHS structure forces us to consider the effects of noise and statistical fluctuations on its evolution. Our Hamiltonian is now $(\int V(r)dr = V_0 \delta\Omega)$

$$H \simeq H_2 = \frac{V_0 \delta\Omega}{N}(\langle\rho\rangle + \delta\rho)\sum_{n=1}^{N}\sigma_1^{(n)}, \qquad (5.125)$$

and we can identify $|n, N - n)$, the symmetrized state in which n molecules are in the ground (unexcited) state, with $|n, N - n\rangle$, defined in (5.112).

According to the analysis here, the coupling constant g turns out to be

$$g = \frac{V_0 \delta\Omega}{N}(\langle\rho\rangle + \delta\rho) \equiv g_0 + \delta g. \qquad (5.126)$$

Therefore, at constant $\langle\rho\rangle$, g is inversely proportional to N. (Note that the average density $\langle\rho\rangle$ is kept constant, when taking the thermodynamical limit.) Since Cini's analysis can be applied identically to our case, this implies that the time for complete discharge of the detector is given by $t_0 = \pi\hbar/2g \propto N$, in this case.

Let us now sketch the behaviour of the system when the detector in a highly excited state approaches a phase transition. The density matrix of the Q+D system is written as

$$\Xi = |\psi\rangle\langle\psi| = \Xi_{\text{diag}} + \Xi_{\text{off}} \longrightarrow \Xi^{\text{F}} = \Xi_{\text{diag}}^{\text{F}} + \Xi_{\text{off}}^{\text{F}}, \qquad (5.127)$$

where the diagonal and off-diagonal terms are respectively given by

$$\Xi_{\text{diag}} = |c_+|^2\eta_{++}\sum_{n=0}^{N}P_n|n, N - n\rangle\langle n, N - n| + |c_-|^2\eta_{--}|N, 0\rangle\langle N, 0|,$$

$$\Xi_{\text{off}} = |c_+|^2\eta_{++}\sum_{n \neq m}a_n a_m^*|n, N - n\rangle\langle m, N - m|$$

$$+ c_+ c_-^* \eta_{+-}\sum_{n=0}^{N}a_n|n, N - n\rangle\langle N, 0| + \text{h.c.}, \qquad (5.128)$$

$\eta_{\pm\pm} = u_\pm u_\pm^\dagger|\phi_\pm|^2$, $\eta_{+-} = u_+ u_-^\dagger\phi_+\phi_-^*$, and Ξ^{F} is the final state. Observe that there are two types of off-diagonal terms, quadratic $(a_n a_m^*)$ and linear (a_n). The former reflect only the coherence among different states of the detector [they would be present even if the initial state of the incoming particle had been $c_+ u_+ \phi_+$, instead of χ_Q, in Eq. (5.110)], while the latter reflect the coherence between different states of the particle as well.

It is possible to prove[135] that in the large-N limit the distribution sought is almost flat, and is characterized by the average

$$\langle n\rangle_{\text{av}} \equiv \int_0^N n\langle P_n\rangle dn \simeq N/2, \qquad (5.129)$$

and the standard deviation

$$\sigma = \left(\int_0^N (n - \langle n \rangle_{av})^2 \langle P_n \rangle dn \right)^{1/2} = 2^{-3/2} N. \qquad (5.130)$$

This result is rather peculiar, and is a consequence of the approximation $H_2 \gg H_Q, H_D$, in (5.121) and (5.122): To neglect H_D means to consider the limit $\hbar\omega \to 0$, i.e. the limit of vanishing energy difference between the excited and ground states of the detector's elementary constituents. Had such a limit not been taken, one should have expected the detector's state to evolve towards the minimum energy state, i.e. the initial state $|N, 0\rangle$. But in our case *all of the detector states are energetically equivalent*, and the resulting distribution is almost flat.

Moreover, a straightforward analysis proves that it is *not* possible to state

$$\Xi \xrightarrow{N \to \infty} \Xi^F_{diag}, \qquad (5.131)$$

because Ξ^F_{off} is not negligible, when compared to Ξ^F_{diag}, even in the $N \to \infty$ limit. To see this, we must estimate the value of

$$\text{Tr}\left(\Xi^F\right)^2 = |c_+|^4 \sum_{n,m=0}^N |\langle a_n a_m^* \rangle|^2 + 2|c_+ c_-^*|^2 \sum_{n=0}^N |\langle a_n \rangle|^2 + |c_-|^4, \qquad (5.132)$$

where the trace is taken over both D and Q. If this is found to be less than 1, we can conclude that the density matrix is partially mixed and coherence has been lost to an extent estimated by the contribution of the off-diagonal matrix elements to the value of the trace. Incidentally, observe that if no average $\langle \cdots \rangle$ were present, one would get

$$\text{Tr}\left(\Xi^F\right)^2 = |c_+|^4 \sum_{n,m=0}^N |a_n|^2 |a_m|^2 + 2|c_+ c_-^*|^2 \sum_{n=0}^N |a_n|^2 + |c_-|^4$$

$$= |c_+|^4 + 2|c_+ c_-^*|^2 + |c_-|^4 = (|c_+|^2 + |c_-|^2)^2 = 1, \qquad (5.133)$$

as expected for a pure state. On the other hand, one can prove[135] that

$$\text{Tr}\left(\Xi^F\right)^2 \sim |c_+|^4 \sqrt{\frac{2}{\pi N}} + 2|c_+ c_-^*|^2 \sqrt{\frac{8}{\pi N}} + |c_-|^4$$

$$\xrightarrow{N \to \infty} |c_-|^4 < |c_-|^2 < 1, \qquad (5.134)$$

as was to be expected for a completely mixed state.[d] This clearly demonstrates that the coherence is lost like $1/\sqrt{N}$ when N becomes large. In the model here

---

[d] A density matrix ρ is called mixed if $\rho^2 \neq \rho$ and in such a case $\text{Tr}\rho^2 < 1$. We may divide density matrices into two classes, corresponding to "partially mixed" and "completely mixed" states, according to the presence and absence of their off-diagonal elements. It is important to note that in this classification, a specific basis has been chosen.

proposed, and in the sense of Eq. (5.134), we can state that in the $N \to \infty$ limit, coherence is lost among different detector states as well as between the two branch waves of the Q particle.

The interpretation of the above results is easily given in terms of the MHS approach: The ergodic assumption (4.25)

$$\overline{\cdots} = \langle \cdots \rangle, \tag{5.135}$$

where the bar denotes, as usual, the average over different incoming particles in an experimental run, yields a consistent interpretation of the results in this subsection. When the same experiment is performed many times on many Q+D systems, the quantum mechanical coherence is lost at a *statistical level*, during the interaction. The loss of coherence, here characterized by the behaviour of $\mathrm{Tr}\,(\Xi^F)^2$ in the $N \to \infty$ limit, is just a statistical effect over many repetitions of the experiment. Observe also that the phase transition, as described in this model, corresponds to the *discharge* of the counter, not to the collapse of the wave function. The collapse, i.e. the loss of coherence, begins already at $t = 0$, immediately after the interaction has been "switched on." Indeed, the coefficients $a_n(t)$ in Eq. (5.114) start feeling the presence of noise, via the coupling constant g, as soon as the evolution governed by the Hamiltonian H_2 begins. The loss of coherence engendered by this process is *partial*, but *unrecoverable*, and we can say that a *partial collapse of the wave function* begins as soon as the microsystem enters into contact with the (macroscopic) detector. The final discharge of D, at $t = t_0$, is therefore a different effect that follows a process started long before. Therefore, as repeatedly emphasized in this book, a clear distinction must be made between the discharge of the counter (the "signal" or in general the "display" of the result of the experiment) and the collapse of the wave function (the "decoherence" of the quantum system, i.e. the transition from pure to mixed state).

Chapter 6

NEUTRON INTERFEROMETRY

Neutron interferometry[30,13] is a very powerful tool not only to measure very small physical quantities but also to investigate the fundamental principles of quantum mechanics. Many experiments performed during the last two decades are very interesting from the measurement-theoretical point of view and enable one to make, in some cases, a critical comparison among alternative theories. In this chapter we shall present and discuss some beautiful neutron interferometric experiments. We will also analyze the experimental data in terms of the decoherence parameter, introduced in Chapter 4, along the line of thought of the MHS approach.

We have to mention that many photon interference experiments have also been performed in connection with fundamental problems in quantum mechanics. We shall not describe such experiments here, for they can be found in many beautiful books and papers on quantum optics.[116,197,119]

Excellent review papers on neutron interferometry, covering experiments as well as the theory of neutron optics, are Refs. 29, 163, 201, 70, 106.

1 Neutron interference

A typical neutron interferometer is made up of a perfect monolithic silicon crystal with three slabs. The atomic planes are parallel throughout the entire crystal, to an excellent approximation, and the crystal is cut in such a way that the Bragg reflecting planes, for an incoming neutron beam, are perpendicular to the face of a slab. The beam is split at the first slab, reflected and redirected at the second slab and finally superposed at the third and last slab. A triangular wedge can be placed, say, along the second route, as in Fig. 6.1: Such a wedge (that can be made of aluminium) acts as a phase shifter (PS) on the neutron state it interacts with. The whole process is *coherent* and, by placing a detector in the ordinary channel O, one can observe a beautiful interference pattern by varying the phase difference δ (proportional to the thickness of the PS) between the two branch waves. A neutron interferometer of this type was first operated by Rauch, Treiner and Bonse.[165]

In the light of the MHS approach, a phase shifter is an excellent example of macroscopic device that does not destroy the quantum coherence: This is readily explained on the basis of the criterion (4.46) (which is equivalent to $\epsilon = 1$) for the loss of quantum coherence: Indeed, the S-matrix representing the interaction of the neutron with the PS yields a phase shift δ that can be simply written as $\delta = -\kappa_0 l$, where l is the (macroscopic) size of the PS, and $\kappa_0 = 2\pi/D_\lambda$, where $D_\lambda = 2\pi/\lambda n b$, λ, n and b being the neutron wavelength, the density of Al atoms in the PS and the scattering length of the low-energy neutron-Al collision, respectively. If we insert the typical numerical values $\lambda \simeq 1.8 \times 10^{-8}$ cm, $n \simeq 6.02 \times 10^{22}$ cm^{-3} and $b \simeq 3.5 \times 10^{-13}$ cm, we obtain $D_\lambda \simeq 1.62 \times 10^{-2}$ cm, which yields

$$\Delta l \ll D_\lambda \qquad \text{or} \qquad \kappa_0 \Delta l \ll 1, \tag{6.1}$$

where Δl is the uncertainty in l. This is opposite to (4.46) and implies $\epsilon \simeq 0$: coherence is preserved between the two branch waves. This confirms the leading idea of the MHS approach: if the statistical fluctuations are very small, a macroscopic device does not provoke any dephasing process.

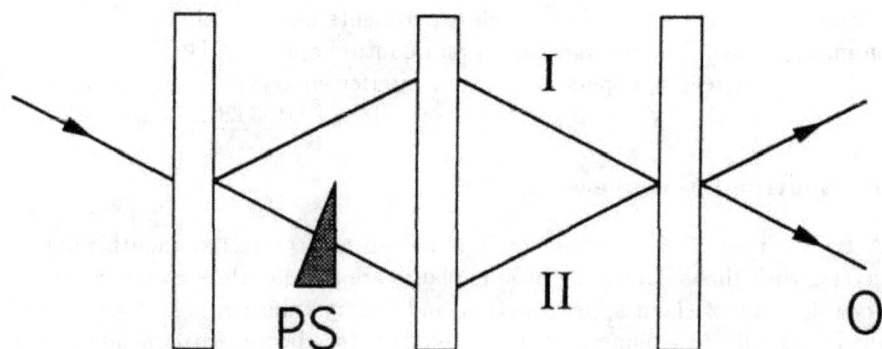

Fig. 6.1. Neutron interferometer. O stands for the ordinary channel.

1.1   Spin-state superposition

Some of the most beautiful seminal experiments of neutron interferometry have led to the experimental verification of the quantum superposition of neutron spin states.[184,11] These experiments provided a direct experimental test of the most basic principles of quantum mechanics, in particular of the superposition principle for fermionic spin states. Let us first describe how a neutron spin flipping device works and then discuss spin-state superposition.

A neutron beam enters a region of space with a static magnetic field B_0 and is energetically separated into two polarized subbeams, whose potential energy difference is $2\mu B_0$, μ being the (modulus of the) magnetic moment of the neutron. Notice that, since the total energy is preserved, this change in potential energy is compensated by a corresponding change in kinetic energy. When the neutron leaves the region of the magnetic field, an opposite change occurs and the neutron undergoes no net energetic effect. However, if a time-dependent magnetic field B_1, perpendicular to the static field B_0, is applied before the neutron leaves the B_0-region, the neutron spin is inverted and the kinetic energy of the outgoing neutron is changed by an amount $4\mu B_0$. This energy is supplied by the electromagnetic field of the rf-coil engendering the B_1 field. See Fig. 6.2. This neutron "magnetic resonance device" was realized by

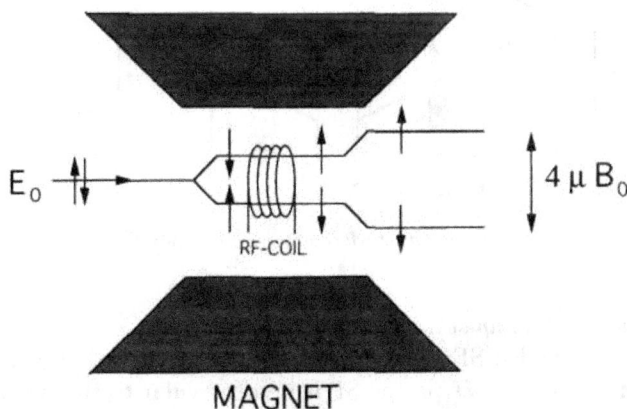

Fig. 6.2. Neutron magnetic resonance device.

Alefeld *et al.* in 1981[1] and had been previously discussed by other authors.[51,12]

The spin-flip probability of a neutron in a time-dependent magnetic field reads[5,98]

$$P_\omega(t) = \frac{\sin^2\left[(\mu B_1 t/2\hbar)\sqrt{1 + [(2\mu B_0 - \hbar\omega)/\mu B_1]^2}\right]}{1 + [(2\mu B_0 - \hbar\omega)/\mu B_1]^2}, \tag{6.2}$$

where t is the time of flight of the neutron through the coil ($t = \ell/v$, ℓ being the length of the coil and v the neutron speed) and ω the frequency of the rf field. If the "resonance conditions"

$$\omega = \omega_L = \frac{2\mu B_0}{\hbar}, \qquad \frac{\mu B_1 \ell}{\hbar v} = \pi \tag{6.3}$$

are met,[1] the spin of the neutron is flipped with probability $P = 1$.

The device just described is a very efficient spin flipper (SF) and can be utilized to check very basic properties of the Schrödinger equation. See Fig. 6.3. We send a neutron beam, polarized along the direction of the static

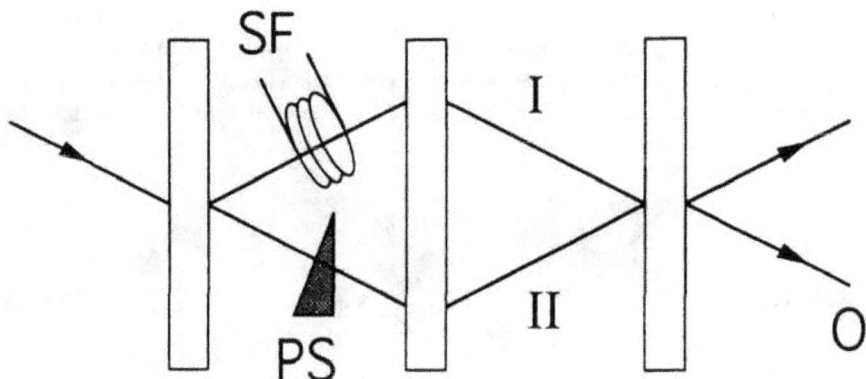

Fig. 6.3. Neutron interferometer with a spin flipper.

magnetic field B_0 (i.e. *up* state), into the interferometer and place a PS along the second route and a SF along the first route of the interferometer. The oscillating magnetic field B_1 in the SF is perpendicular to the static field B_0.

The neutron states superposed at the third slab of the interferometer have orthogonal spins and do not interfere. However, they are perfectly coherent, because all interactions (included those at PS and SF) preserve the quantum coherence. The Vienna group invented a sophisticated device[11,184] in order to bring to light the coherence between the *up*-polarized component coming from the second route and the *down*-polarized one coming from the first route. Since this experiment is of particular interest from the measurement-theoretical point of view, it is worth going into more details.

Let

$$\psi_I = \frac{1}{\sqrt{2}}(\phi_1 + \phi_2)|\uparrow\rangle \tag{6.4}$$

be the initial state just before the PS and SF, where $|\uparrow\rangle$ is the spin-up state along the direction of $\boldsymbol{B}_0$ (say z) and $\phi_1(\phi_2)$ represents the wave packet propagating in the first (second) route. The action of PS and SF, under the resonance condition (6.3), yields the final state (ordinary channel)

$$\psi_{\mathrm{F}} = \frac{1}{\sqrt{2}}(-i\phi_1|\uparrow\rangle + e^{i\delta}\phi_2|\downarrow\rangle). \tag{6.5}$$

The polarization in the O channel

$$\boldsymbol{P}_{\mathrm{O}} = \langle\psi_{\mathrm{F}}|\boldsymbol{\sigma}|\psi_{\mathrm{F}}\rangle = (-\sin\delta, \cos\delta, 0) \tag{6.6}$$

lies in the x-y plane (remember that the initial polarization was $+z$ and notice that the state ψ_{F} in (6.5) is a superposition of $\pm z$ states). The Vienna group[184] placed in the O channel a so-called $\pi/2$-spin turn coil, which interchanges the y- and z-components of the polarization vector, in front of a magnetic Heusler single crystal analyzer aligned in the z-direction: a beautiful interference pattern was observed as a function of δ. The experiment was also performed in a time-dependent rf field $\boldsymbol{B}_1(t)$, rotating in the x-y plane.[11] A calculation analogous to the one shown above yields a time-dependent polarization in the O channel

$$\boldsymbol{P}_{\mathrm{O}}(t) = (-\sin[\delta - \omega t], \cos[\delta - \omega t], 0). \tag{6.7}$$

This was measured by means of a stroboscopic technique.

The experiments just described bring up an interesting question, from the measurement-theoretical point of view. One may think that since neutron spin-flip takes place, in these experiments, the neutron route is determined. If this were true then, according to the naive Copenhagen interpretation, one should conclude that one of the two branch waves must have disappeared, and therefore no interference should be observed. On the contrary, the experiment clearly showed an interference pattern. The experimental result was understood on the basis of the number-phase uncertainty relation within the framework of quantum mechanics.[1,11]

Vigier and coworkers[48,194] asserted that the experimental result was in conflict with the Copenhagen interpretation, that regards a quantum object as a wave *or* a particle, and supported their theory that regards it as a wave *and* a particle. They also proposed another experiment in which two spin flippers were placed along both routes of the interferometer. Interference can then be observed without making use of stroboscopic techniques and no arguments based on the phase-number uncertainty relation for fermions are needed. A new experiment was performed, yielding again a beautiful interference pattern.[1,11]

It seems to us that the claim in Refs. 48, 194 was a trifle hasty. A dynamical process like spin flipping must be strictly distinguished from the determination of the particle path. Let us analyze the above-mentioned experiment from the viewpoint of the MHS approach.[140] By making use of the formula

$$\frac{l}{v}\frac{\omega_L}{2}\sqrt{1 + (B_1/B_0)^2} = \frac{2\pi l}{D_\lambda},\tag{6.8}$$

that expresses the neutron phase shift provoked by an oscillating magnetic field of frequency ω_L if the frequency resonance condition (6.3) is met, one obtains, for $B_1 \gg B_0$, $l \simeq 1.5$ cm, and $v \simeq 2.2 \times 10^5$cm·s^{-1}, the value $D_\lambda \simeq 10$ cm. Of course

$$\Delta l \ll l = 1.5 \text{ cm} < D_\lambda \simeq 10 \text{ cm},\tag{6.9}$$

and the condition (6.1), yielding $\epsilon \simeq 0$, is again satisfied. Even though spin flipping occurs in one of the two paths, this does *not* provoke a dephasing process and interference can still be observed. Spin flipping in one of the two routes of the interferometer does *not* imply any determination of the neutron trajectory. In conclusion, the MHS approach predicts that coherence is kept between the branch waves and no wave function collapse takes place. The experiment is therefore not in conflict with the Copenhagen interpretation that regards a quantum object as a wave *or* a particle. Notice that this discussion does not rely upon the number-phase uncertainty relation.

We conclude this survey by commenting on another neutron interference experiment, in which small pulses are forwarded to the interferometer, in such a way that one pulse contains on the average 0.003 neutrons.[73] Each pulse is divided into two branch waves, and interference is observed after recombination in the final channel. One can roughly state that the two leading parts, the two central bodies, and the two trailing ends of the two branch waves of the same wave packet give rise to interference, even though this pictorial view is not unanimously accepted.[42,111]

The experiment seems to suggest that a certain "reality" can be ascribed to the Schrödinger wave function, in some sense: It is a widespread opinion, among many physicists, that the Schrödinger wave function is to be regarded only as a mathematical tool, but does not correspond to any physical reality. However, the above-mentioned experiment seems to suggest that the Schrödinger wave function can have a certain reality in some sense, because a very small fraction ($\simeq 0.003$) of a single-particle wave function behaves as if it could interact with an apparatus separately from its other fractions. The situation is very similar to the case of classical waves, in which a very small fraction interacts with the surrounding materials, independently from other fractions, and produces the corresponding physical results. Of course, however, we know that an "interaction" in the quantum case does not always yield a "result," as was described in Chapter 2. In other words, the Schrödinger wave function remarkably differs from classical waves, mainly because of the wave-particle dualism and the probabilistic interpretation. For this reason we are led to wonder up to which extent the above-mentioned experimental fact seems to favor the rather strong viewpoint that the Schrödinger wave function has got a certain reality, in contrast with the widespread opinion that it is only a mathematical tool. The final answer to this question will be given in the future: Quantum mechanics is still a developing theory and does not seem to have been completely established. Anyway, we remark that this experiment is interesting, for it shows that a Schrödinger wave function for a massive particle is able to interfere even though there is "less than one particle" in the apparatus, at any given time.

1.2 Gravity experiments on neutron and photon

One of the most spectacular experiments of neutron interferometry is that evidentiating the presence of a gravitational phase shift. The seminal experiment is known as the COW experiment, after the initials of Colella, Overhauser and Werner.[41,150] Interesting papers on this subject, focusing on the conceptual difficulties (some of which still remain) are Refs. 70, 180.

In this experiment, the interferometer sketched in Fig. 6.1 is rotated by an angle α so that one of the routes, say II, is higher than the other. The difference in gravitational potential energy along the two routes is reflected in the phase difference

$$\Delta\phi = \frac{mgA\sin\alpha}{\hbar v}, \qquad (6.10)$$

where m is the neutron mass, g the gravitational acceleration, A the area enclosed between the two routes and v the neutron speed. This phase shift was experimentally observed as a function of α.

This spectacular experiment is only possible due to the rather large value of the neutron mass $m \simeq 10^{-24}$g. No analogous experiment is feasible with electrons or photons.

If we replace the neutron in the above gravitational experiment with a photon, we obtain a photon interference experiment in the Earth's gravitational field. Such an experiment would be very interesting and ambitious, because it would enable us to examine an interferometric general-relativistic effect on the photon. Usually, a possible photon phase shift is anticipated to be too small to be detected, because the photon gravitational mass and the area enclosed between the two photon paths is very small. However, future experimental progress will overcome such practical difficulties, by preparing photons with very high frequency and a long photon path. Actually, the Japanese Spring8 group is planning to perform this kind of experiment.

1.3   Coherence and postselection

Neutron interference experiments are always performed with beams coming out of a reactor and having a certain spread. The first measurements of the so-called longitudinal coherence length of a neutron wave packet were performed simultaneously by Kaiser *et al.*[94] and Klein *et al.*[105] Very accurate estimates of the three-dimensional coherence functions were obtained by Rauch *et al.*[166]

An interference pattern is always observed by placing a phase shifter (PS) in one of the two routes of the interferometer. The PS yields a phase shift as well as a spatial shift on the neutron wave function, so that the longitudinal coherence length limits the fringe visibility of the interference pattern.

An interesting idea aimed at restoring the loss of interference contrast due to the finite coherence length was put forward by Clothier *et al.*[40] The fringe visibility is partially restored by placing after the PS, in the same arm of the interferometer, another slab of material with a negative optical potential. This phenomenon was named "phase echo." A much more spectacular result was obtained later by the same authors:[93,87] By making use of a *postselection* technique, it is possible to restore the fringe visibility after the neutrons have come out of the interferometer. Let us discuss this beautiful idea. A review can be found in Ref. 162.

Let Δ be the spatial shift provoked by the PS, placed in the second route of the interferometer. $\Delta(k)$ is a function of the wave vector k and is proportional to the thickness of the PS. The so-called autocorrelation function[87,162] is nothing but the overlap function between two wave-packet states and is defined by the convolution

$$\Gamma(\Delta) = \langle \psi(0)|\psi(\Delta)\rangle, \tag{6.11}$$

where $|\psi(\Delta)\rangle$ stands for a wave-packet state with its phase shifted by the amount Δ by the PS. The intensity in the ordinary channel is given by

$$I_O \propto 1 + |\Gamma(\Delta_0)| \cos\delta_0, \tag{6.12}$$

where $\Delta_0 = \Delta(k_0) = (\Delta_x, \Delta_y, \Delta_z)$ and $\delta_0 = \Delta_0 \cdot k_0$ is the phase shift, k_0 being the average wave vector of the wave packet. For a Gaussian wave packet, the modulus of Γ is readily calculated to be

$$|\Gamma(\Delta_0)| = \prod_{i=x,y,z} \exp[-(\Delta_i \delta k_i)^2/2], \tag{6.13}$$

where δk_i is the characteristic spread of the momentum distribution of the wave packet along direction i ($i = x, y, z$) and $\Delta(k)$ has been assumed to be a slowly varying function of k. Notice that the above relation defines the coherence lengths $\Delta_i^c = 1/2\delta k_i$ of the wave packet, according to the uncertainty relations. Equation (6.12) is readily understood in terms of the *spatial overlap* of the two branch waves: If the PS is thick, so that $|\Delta_0|$ is large, the branch waves do not overlap and interference is lost.

However, the PS is an object that preserves the coherence of the neutron, so that it is possible to bring such a coherence to light. The Missouri-Vienna group[93,87] placed a momentum analyzer crystal in the ordinary beam in order to retrieve the coherence information "stored" in a particular momentum component of the outgoing state. The intensity for the selected momentum component k in the ordinary channel is easily computed:

$$I_O(k) \propto |\widetilde{\psi}(k)|^2[1 + \cos(\Delta(k) \cdot k)]^2, \tag{6.14}$$

where $\widetilde{\psi}$ is the Fourier transform of the wave packet. If $|\Delta|$ is large, even though, according to (6.12) and (6.13), interference disappears for the overall beam, an interference pattern is clearly obtained by selecting specific momentum components, as predicted by (6.14). Notice that the interfering momentum components are selected *after* the two branch waves have been recombined in the final (ordinary) channel. For these reasons the experimental technique was named "postselection."

These experiments shed light on an important principle: The lack of spatial overlap between the interfering wave functions implies a lack of interference, but does not imply at all a loss of quantum mechanical coherence. No overlap does not necessarily mean no coherence, as was repeatedly explained in this book (Chapter 4, in particular). The quantum coherence can always be brought to light if the experimenter is able to invent a suitable device, like in the case just described. This result is in line with the the philosophy of the MHS approach. See, in particular, the discussion in Section *2.1* of Chapter 3 and in Section *1.2* of Chapter 4.

2 Neutron interference at low transmission probability

Let us now discuss a very interesting experiment performed a few years ago by the Vienna group.[185,161,164] The experimental data showed a discrepancy from the theoretical prediction for the visibility: The experimental points lay remarkably below the quantum mechanical curve, which is proportional to $\sqrt{t}$ in the low t region, t being the transmission probability along one route of the interferometer. The reduction of the interference term in this experiment was ascribed[141,145,128] to the presence of statistical fluctuations of the elementary constituents which absorb neutrons and the result was understood along the line of thought of imperfect measurements. This interpretation was corroborated by an evaluation of the decoherence parameter. We start by analyzing a neutron interferometric experiment when an absorber is placed in one of the two routes of the interferometer. We then show how the decoherence parameter emerges in a natural way when the statistical fluctuations of the absorber are considered.

2.1 *Phenomenological treatment*

Let us first show, by means of a heuristic argument,[145] that the reduction of the visibility in a neutron interferometric experiment is a physical consequence of the presence of statistical fluctuations in the absorber.

Let the incident neutron wave packet be split into two branch waves ψ_1 and ψ_2, corresponding to the two different routes in the apparatus, and assume that ψ_2 interacts first with a phase shifter PS and then with an absorber A. The first contributes a phase factor $e^{i\delta}$ while the second is assumed to simply multiply the wave function by a transmission coefficient T, so that

$$\psi_2 \longrightarrow e^{i\delta} T \psi_2. \qquad (6.15)$$

If ψ_1 and ψ_2 are in phase and $|\psi_1|^2 = |\psi_2|^2$ (at some spatial point in the final channel where the two waves are recombined), the intensity after recombination of the two branch waves is

$$I \propto |\psi_1 + e^{i\delta} T \psi_2|^2 \propto 1 + |T|^2 + 2\mathrm{Re}(Te^{i\delta}) = 1 + t + 2\sqrt{t}\cos(\alpha + \delta), \quad (6.16)$$

where we have written $T = |T|e^{i\alpha}$, and have defined the transmission probability $t = |T|^2$. In this way, the visibility of the interference pattern is given by

$$V_0 = \frac{I_{\mathrm{MAX}} - I_{\min}}{I_{\mathrm{MAX}} + I_{\min}} = \frac{2\sqrt{t}}{1 + t}. \qquad (6.17)$$

Notice that in the above formulae the internal dynamics of the macroscopic apparata has been ignored, and the effect of their interaction with the neutron wave function has been "summarized" by introducing the two "constants" δ and T in Eq. (6.15). Obviously, this is only an approximation, because both the phase shifter and the absorber are macrosystems, made up of a huge number of elementary constituents and characterized by a few macroscopic parameters whose values cannot precisely determine the details of the microscopic motion. For this reason, fluctuations and uncertainties should be taken into account whenever we consider an interaction between a quantum system and a macrosystem. Our first purpose is to analyze the validity of the approximation (6.15): This will be done in the spirit of the MHS approach.

First of all, observe that the graph of an interference pattern is made up of a certain number of experimental points, each of which represents an intensity (number of neutrons hitting a detector per unit time) versus a value of the phase δ. Each experimental point in the graph is obtained by accumulating the results relative to a very large number of neutrons, that are sent into the interferometer through a weak and steady beam. A given point represents the intensity detected in one of the two channels (say the ordinary one), in relation to a "precise" value of the phase δ acquired by each neutron after the interaction with the PS. This is obviously a *very reasonable* approximation: Indeed, a "good" phase shifter must yield a constant phase factor for every neutron in the same experimental run. Were this factor not "constant" (up to a very good approximation), the interference experiment itself would be impossible to perform. See the criterion (6.1).

On the other hand, the validity of the assumption (6.15) is less obvious for the transmission coefficient T. Indeed, the fluctuations of T, unlike those of δ, can be important. This is due to two main reasons. First, the thickness D of the absorber cannot be considered constant, from event to event, because of the sample inhomogeneities and of the angular divergence of the beam. Second, as repeatedly stressed in this book, in an interference experiment one must accumulate a huge number of experimental results produced by neutrons sent through a very weak and steady beam, in order to obtain an interference pattern. Even though, during an experimental run, the *macroscopic* state of the absorber is always the same, each neutron will interact with a slightly different *microscopic* state of it: Indeed, the atoms responsible for neutron absorption are subject to their own internal motion, and their positions change all the time; moreover, different neutrons will go through (and interact with) different parts of the absorber, due to the finite lateral size of the beam.

Let us now estimate the value of the decoherence parameter ϵ and show that it plays an important role in the analysis of the above-mentioned experiments.[185,161,164] We label different incoming neutrons with j ($j = 1, \ldots, N_p$, where N_p is the total number of neutrons in an experimental run) and compute the average intensity (6.16), by making use of the ergodic assumption (4.25). The visibility will be given by Eq. (4.16), where we shall obviously identify $\overline{|T|^2} = \langle |T|^2 \rangle$ with t, the *experimentally measured* value of the transmission probability.

We start from Goldberger's formula[68,174]

$$T_n = \exp\left[-\left(i\lambda b_R + \frac{1}{2}\sigma_a\right)\rho D\right] = \exp\left[-\frac{1}{2}n\left(1 + i\frac{2\lambda b_R}{\sigma_a}\right)\right], \qquad (6.18)$$

where λ is the wavelength of the neutron, b_R the real part of the scattering length of the elastic neutron-Gd collision, σ_a the absorption cross section for neutron-Gd, ρ the density of (Gd) scatterers, and D the thickness of the absorber A. The quantity $n = \sigma_a \rho D$ is interpreted as the number of scatterers met by the neutron during its interaction with A. In the Vienna experiments, A consisted of a Gd-H_2O solution, so that the neutrons were mostly absorbed by the Gd atoms.

Equation (6.18) is easily understood by heuristically assuming that, as far as absorption and transmission probabilities are concerned, each neutron interacts with a small cylinder of Gd-H_2O solution: This cylinder has height roughly equal to the length of the absorber, and base roughly equal to the neutron-Gd absorption cross section: $\ell = (\sigma_a \rho)^{-1}$ is the mean free path of a neutron for absorptive scattering by Gd atoms. We are neglecting the role of water in the process, because water does not strongly absorb neutrons.

In the spirit of the MHS approach, we assume that n (the number of elementary constituents met by every neutron) is not constant, but rather fluctuates around its average

$$n = \langle n \rangle + \delta n, \tag{6.19}$$

and require the Gaussian properties:

$$\langle \delta n \rangle = 0, \qquad \langle (\delta n)^2 \rangle = g \langle n \rangle, \tag{6.20}$$

where g ($0 \leq g \leq 1$) is a parameter characterizing the fluctuation: The limiting cases $g = 0$ and $g = 1$, correspond to absence of fluctuations and Poissonian fluctuations, respectively. The latter case represents the limit of the dilute solution, or alternatively an ideal-gas correlation function.[141] The parameter g represents the strength of the fluctuations, or alternatively, the size of the uncertainties in some macroscopic parameter, such as D. We are neglecting the density fluctuation of water. This assumption is sound, because water contributes an almost constant factor in the transmission coefficient.

By using the Gaussian reduction formula, we obtain

$$\overline{T} = \langle T_n \rangle = T_0 \exp \left[\frac{1}{8} g \langle n \rangle \left(1 + i \frac{2\lambda b_R}{\sigma_a} \right)^2 \right], \tag{6.21}$$

where

$$T_0 \equiv \exp \left[-\frac{1}{2} \langle n \rangle \left(1 + i \frac{2\lambda b_R}{\sigma_a} \right) \right], \tag{6.22}$$

and

$$t \equiv \langle |T_n|^2 \rangle = t_0 \exp \left[\frac{1}{2} g \langle n \rangle \right], \qquad t_0 = |T_0|^2. \tag{6.23}$$

From Eqs. (6.23) and (6.21) we readily obtain

$$|\overline{T}|^2 = t^{1+\gamma}, \qquad \gamma = \frac{g/4}{1 - g/2} \frac{(\lambda b_R)^2 + (\sigma_a/2)^2}{(\sigma_a/2)^2} \tag{6.24}$$

and the visibility can be rewritten in terms of γ as

$$V_{\text{MHS}} = \frac{2 t^{\frac{1+\gamma}{2}}}{1 + t} = V_0 t^{\frac{\gamma}{2}}, \tag{6.25}$$

where V_0 is defined in (6.17). It is worth stressing that the above equation is liable to *direct* experimental check: By inferring the value of g from Eq. (6.23), we can test the validity of Eq. (6.25).

Let us briefly discuss the main consequences of our analysis. We have shown that, if uncertainties and fluctuations are taken into account, the usual relation between transmission coefficient and probability ($|T|^2 = t$) is not valid anymore, and must be replaced by Eq. (6.24). Accordingly, the value of the visibility is reduced by a factor $t^{\frac{1}{2}}$, as shown in (6.25). This suggests how the effect can be checked experimentally: Indeed, the correction to the visibility (6.25) is negligible when $t \simeq 1$, but becomes very important when $t \to 0$. This is the reason why the reduction of the visibility becomes important at extremely low values of the transmission probability. See also the discussion in Section *1.4* of Chapter 7 and Fig. 7.10 in particular.

It is very interesting to discuss the results obtained from a measurement-theoretical point of view. The decoherence parameter is

$$\epsilon = 1 - \exp\left[-\frac{g\langle n\rangle}{4}\frac{(\lambda b_R)^2 + (\sigma_a/2)^2}{(\sigma_a/2)^2}\right] = \epsilon(g, \langle n\rangle) \tag{6.26}$$

and can be expressed in terms of t and γ as

$$\epsilon = 1 - t^\gamma. \tag{6.27}$$

Analogously to the case of the visibility, discussed before, this implies that even though, at high transmission probability ($t \simeq 1$) fluctuation effects are not observable, they become very important when $t \to 0$. In such a case, $\epsilon \to 1$, and quantum coherence is totally lost. Observe that this effect is completely independent of the fact that one of the two branch waves is (almost) totally absorbed: Indeed, even if t is extremely small (say, of order 10^{-5}), both branch waves are still *present* in the interferometer, and always give rise to interference. The point is that this interference is drastically reduced with respect to the value (6.17). In this sense, one can state that a perfect preservation of coherence is *impossible* even *in principle*. This idea will be corroborated by numerical results in the following chapter. In particular, we will see in Section *1.4* of the following chapter that the decoherence parameter takes values close to one in the region of very low transmission probability. This suggests that the quantum coherence cannot be preserved when the probability of absorption becomes very large. This is an important point, from an epistemological point of view, and is extraneous to von Neumann's theory and to the standard Copenhagen interpretation. Indeed, observe that the effect we are pointing out, and in general any dephasing effects describable by the decoherence parameter ϵ, can be analyzed only if the microscopic structure of the macrosystem is taken into account.

A more exhaustive analysis,[128] that takes the presence of water into account and neglects the effects of the uncertainties in D in comparison with those provoked by the Gd-density fluctuations, yields the following formulae for a neutron with wave number k, passing through an absorber of thickness D, containing a solution of Gd in water

$$\epsilon = 1 - \exp\left[-\frac{\pi}{k^2}\frac{\sigma_s^G}{\sigma_t^G}g\langle\rho_G\rangle D\right],\tag{6.28}$$

$$\gamma = \frac{\pi g\sigma_s^G/k^2\sigma_t^G\sigma_a^G}{\left[1 - (g\sigma_a^G/2\sigma_t^G) + (\sigma_a^W\rho_W/\sigma_a^G\langle\rho_G\rangle)\right]}.\tag{6.29}$$

Here σ_a^G (σ_a^W) is the absorption cross section for neutron-Gd (neutron-water), σ_t^G the total cross section for a neutron-Gd collision and σ_s^G the cross section for an elastic neutron-Gd collision,

$$\sigma_s^G = 4\pi\left[\left(b_R^G\right)^2 + \left(\frac{k^2\sigma_a^G}{4\pi}\right)^2\right],\tag{6.30}$$

where b_R^G and σ_a^G are the real part of the scattering length and the absorption cross section for an elastic neutron-Gd collision, respectively, ρ_G and ρ_W the Gd and water densities, respectively, expressed in number of particles per unit volume, and for simplicity we assumed ρ_W constant.

In the above formulae, g represents the strength of the correlation and is given by

$$g \equiv 2\sigma_t^G \int_0^D \overline{G}(x)dx,\tag{6.31}$$

where

$$\overline{G}(x - x') \equiv \frac{1}{(\sigma_t^G)^2}\int\cdots\int G(x - x', y - y', z - z')dydzdy'dz'\tag{6.32}$$

and we have decomposed the Gd-density function into its average and fluctuation part

$$\rho_G(\boldsymbol{r}) = \langle\rho_G\rangle + \delta\rho_G(\boldsymbol{r})\tag{6.33}$$

and have written

$$\langle\delta\rho_G(\boldsymbol{r})\delta\rho_G(\boldsymbol{r}')\rangle = \langle\rho_G\rangle G(\boldsymbol{r} - \boldsymbol{r}').\tag{6.34}$$

The function $G(\boldsymbol{r} - \boldsymbol{r}')$ represents the density correlation function of Gd atoms in the Gd-H$_2$O solution.

Let us briefly discuss the role of g. If we had just an ideal Gd gas in the absorber, we could set $G(\boldsymbol{r} - \boldsymbol{r}') = \delta(\boldsymbol{r} - \boldsymbol{r}')$ (dilute-solution limit) and obtain $g = 1$. In this case we would overestimate the fluctuations, in comparison with

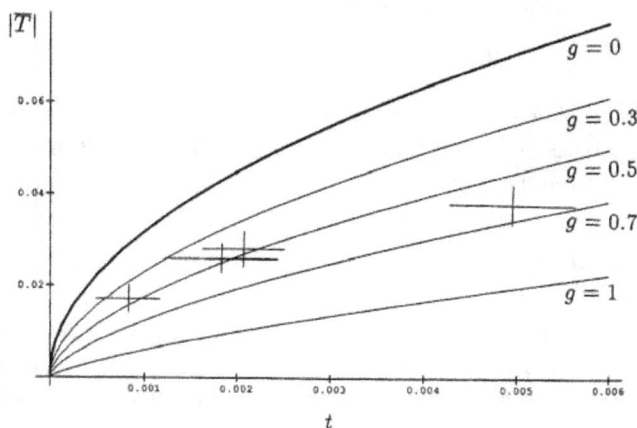

Fig. 6.4. Comparison of the experimental data with the theoretical predictions for different values of the parameter g. The cases $g = 0, 1$ correspond to absence of fluctuations and dilute-solution (or ideal-gas) correlation function, respectively.

a realistic absorber. For this reason we can generally consider g as a reduction factor, smaller than unity.

Figure 6.4 shows the numerical results in the low-t region for the values $g = 0, 0.3, 0.5, 0.7$ and 1, where we took

$$\langle \rho_G \rangle = 5 \times 10^{26} \text{m}^{-3}. \tag{6.35}$$

The experimental points in Fig. 6.4 were obtained by the Vienna group.

2.2 Hamiltonian treatment: Perturbative approach

In the previous subsection we gave a simple but still accurate estimate of the loss of quantum coherence undergone by the neutron beam. We have shown in a nonperturbative way [Eqs. (6.28) and (6.29)] that, with D fixed, the experimentally observed reduction of interference at low transmission probability can be explained by taking into account the density fluctuations in the absorber, on the basis of Goldberger's formula (6.18). However, we should not forget that Eq. (6.18) is an approximate formula, unable to take into account the details of the space-time relaxation or the spectral distribution of the fluctuations. In order to do this, we have to handle the whole neutron+absorber system quantum mechanically and analyze the collision process of an incoming neutron with an ensemble of Gd atoms. Let us outline a possible theoretical approach to this problem, by following Refs. 141, 128.

Our system is described by the Hamiltonian

$$\hat{H} = \hat{H}_0^Q + \hat{H}_0^A + \hat{H}' = \hat{H}_0 + \hat{H}', \qquad (6.36)$$

where $\hat{H}_0^Q$ is the free Hamiltonian of the neutron, $\hat{H}_0^A$ the free Hamiltonian of the absorber, $\hat{H}_0 = \hat{H}_0^Q + \hat{H}_0^A$, and $\hat{H}'$ the interaction Hamiltonian. The latter is explicitly written as

$$\hat{H}' = \sum_{n=1}^{N} \gamma_n V(\hat{r} - r_n) = \int d^3 r' \rho(\hat{r} - r') V(r'), \qquad (6.37)$$

where N is the total number of elementary constituents (Gd atoms) of the absorber, r_n the position of the nth constituent, V the interaction potential between neutron and Gd atom (V is complex, in order to take into account the possibility of neutron absorption) and

$$\rho(\hat{r}) = \sum_{n=1}^{N} \gamma_n \delta(\hat{r} - r_n) \qquad (6.38)$$

is the density of Gd atoms. Here we suppress the subscript G for the density ρ. In general, we can take into account the inner structure of the scatterers (Gd atoms) and the possible change of their inner state after the interaction with the neutron, if we consider an operator-valued γ_n acting on the state of the nth Gd atom. For simplicity, we neglect the internal structure of Gd atoms and assume $\gamma_n = 1$, so that ρ is normalized to N.

Define

$$\hat{\mathcal{H}}_0^Q = \hat{H}_0^Q + \mathcal{V}, \quad \hat{\mathcal{H}}' = \hat{H}' - \mathcal{V}, \qquad (6.39)$$

where $\mathcal{V} = \mathcal{V}_R - i\mathcal{V}_I \in \mathbb{C}$ is a constant "optical" potential. Equation (6.36) can be rewritten as

$$\hat{H} = \hat{\mathcal{H}}_0^Q + \hat{H}_0^A + \hat{\mathcal{H}}' \qquad (6.40)$$

and the S-matrix for the total system is

$$\hat{S} = e^{\frac{i}{\hbar}\hat{H}_0 \tau} e^{-\frac{i}{\hbar}\hat{H}\tau} = S_0 \hat{U}_I(\tau), \qquad (6.41)$$

where τ is the total transit time of the particle in the absorber and we have defined

$$S_0 = e^{\frac{i}{\hbar}\hat{H}_0^Q \tau} e^{-\frac{i}{\hbar}\hat{\mathcal{H}}_0^Q \tau} = e^{-\frac{i}{\hbar}\mathcal{V}\tau}, \qquad \hat{U}_I(\tau) = e^{\frac{i}{\hbar}\hat{\mathcal{H}}_0 \tau} e^{-\frac{i}{\hbar}\hat{H}\tau}, \qquad (6.42)$$

with $\hat{\mathcal{H}}_0 = \hat{\mathcal{H}}_0^Q + \hat{H}_0^A$. It is easy to show that $\hat{U}_I(t)$ obeys the equation

$$i\hbar \frac{d\hat{U}_I(t)}{dt} = \hat{\mathcal{H}}_I'(t)\hat{U}_I(t), \qquad (6.43)$$

where

$$\hat{\mathcal{H}}_1'(t) = e^{\frac{i}{\hbar}\hat{H}_0 t}\hat{\mathcal{H}}'e^{-\frac{i}{\hbar}\hat{H}_0 t} = e^{\frac{i}{\hbar}\hat{H}_0 t}\hat{\mathcal{H}}'e^{-\frac{i}{\hbar}\hat{H}_0 t} \qquad (6.44)$$

and the last equality in (6.44) is due to the fact that $\mathcal{V}$ is constant. Equation (6.43) allows us to write the Dyson series:

$$\hat{S} = S_0 \left(1 + \frac{1}{i\hbar}\int_0^\tau dt \hat{\mathcal{H}}_1'(t) + \frac{1}{(i\hbar)^2}\int_0^\tau dt \int_0^t dt' \hat{\mathcal{H}}_1'(t)\hat{\mathcal{H}}_1'(t') + \dots\right). \qquad (6.45)$$

Due to (6.37), the interaction Hamiltonian in (6.44) is explicitly written as

$$\hat{\mathcal{H}}_1'(t) = \left[\int d^3 r' e^{\frac{i}{\hbar}\hat{H}_0^Q t}\rho(\hat{r} - r', t)e^{-\frac{i}{\hbar}\hat{H}_0^Q t}V(r')\right] - \mathcal{V}, \qquad (6.46)$$

where only the partial evolution over the absorber variables has been carried out:

$$\rho(\hat{r}, t) = e^{\frac{i}{\hbar}\hat{H}_0^A t}\rho(\hat{r})e^{-\frac{i}{\hbar}\hat{H}_0^A t}. \qquad (6.47)$$

To compute $\rho(\hat{r}, t)$ is impossible, in practice, due to the huge number N of elementary constituents of the absorber A. We follow therefore the philosophy of Eq. (6.33) and assume that $\rho(r, t)$, the position representation of the above $\rho(\hat{r}, t)$, is made up of a constant, background density $\langle \rho \rangle$ and a "noise" $\delta\rho$,

$$\rho(r, t) = \langle \rho \rangle + \delta\rho(r, t). \qquad (6.48)$$

We require the noise to satisfy the following statistical properties:

$$\langle \delta\rho(r, t) \rangle = 0, \qquad \langle \delta\rho(r, t)\delta\rho(r', t') \rangle = \langle \rho \rangle G_\theta(r - r', t - t'), \qquad (6.49)$$

where the brackets denote the statistical ensemble average over all the possible absorber states and G_θ is, in general, an increasing function of the temperature θ of the absorber as well as a function of space and time.

By inserting (6.48) in (6.46), we obtain

$$\hat{\mathcal{H}}_1'(t) = \int d^3 r' e^{\frac{i}{\hbar}\hat{H}_0^Q t}\delta\rho(\hat{r} - r', t)e^{-\frac{i}{\hbar}\hat{H}_0^Q t}V(r'), \qquad (6.50)$$

where we have required that $\langle \hat{\mathcal{H}}_1'(t) \rangle = 0$, or equivalently

$$\mathcal{V} = \langle \rho \rangle V_0 \delta\Omega, \qquad (6.51)$$

with $V_0\delta\Omega = \int d^3 r V(r)$ (being mainly interested in low-energy (thermal) neutron collisions, we can use the effective potential $V(r) = V_0\delta\Omega\delta^3(r)$, with $V_0\delta\Omega = (2\pi\hbar^2 b/m)$, b being the scattering length).

Let us now observe that the quantity S_0, defined in (6.42), is a c-number, and can be set equal to

$$S_0 = e^{-\frac{i}{\hbar}V\tau} = T_0 = \sqrt{t_0}e^{i\beta}, \tag{6.52}$$

where

$$t_0 = |T_0|^2 = |S_0|^2 = e^{-\frac{2}{\hbar}V_I\tau}, \qquad \beta = -\frac{1}{\hbar}V_R\tau. \tag{6.53}$$

We now write the neutron transmission coefficient, corresponding to the jth incoming neutron, as

$$T_j = T_0(1 + \Delta_j), \qquad j = 1, \ldots, N_p. \tag{6.54}$$

Here T_0 is an ideal parameter, that can be identified with that in (6.52), and the dependence of the transmission coefficient on j is summarized in Δ_j. The average transmission coefficient will be

$$\overline{T} = T_0(1 + \overline{\Delta}), \tag{6.55}$$

where, as usual, a bar denotes an average over j. Note that $\overline{\Delta}$ does not necessarily vanish: This can be understood by realizing that $\overline{\Delta}$ represents an effect due to the statistical fluctuations in the macroscopic detector, which, in general, do not vanish even at zero temperature.

The average transmission probability is given by

$$t = \overline{|T|^2} = t_0(1 + 2\,\mathrm{Re}\,\overline{\Delta} + \overline{|\Delta|^2}), \tag{6.56}$$

where $t_0 = |T_0|^2$ can be identified with the quantity in (6.53). By combining (6.55) and (6.56) we obtain

$$\epsilon = 1 - \frac{|\overline{T}|^2}{\overline{|T|^2}} = \frac{(\delta\Delta)^2}{|1 + \Delta|^2}, \tag{6.57}$$

so that ϵ is proportional to the square of the standard deviation $(\delta\Delta)^2 = \overline{|\Delta|^2} - |\overline{\Delta}|^2 = \overline{|\Delta - \overline{\Delta}|^2}$.

In general, however, the transmission coefficient is defined by the S-matrix element $\langle\phi_{k'}|\hat{S}|\phi_k\rangle$ between the two plane waves $\phi_{k'}(r) = \exp[ik' \cdot r]/\sqrt{\Omega}$ and $\phi_k(r) = \exp[ik \cdot r]/\sqrt{\Omega}$, Ω being the normalization volume. After the above decomposition procedure of the Hamiltonian and the Dyson expansion for $\langle\phi_{k'}|\hat{U}_\mathrm{I}(\tau)|\phi_k\rangle$, we obtain the first-order approximation

$$\Delta_1 = -\frac{i}{\hbar}\frac{V_0\delta\Omega}{\Omega}\int_0^\tau dt \int d^3r\,\delta\rho(r, t)e^{i(k-k')\cdot r - i(\omega_k - \omega_{k'})t} \tag{6.58}$$

for the potential $V(r) = (V_0\delta\Omega)\delta(r)$, in which $\omega_k = (\hbar k^2/2m)$.

Therefore, the averaged square of the first-order term reads

$$\overline{|\Delta_1|^2} = \frac{\langle\rho\rangle|V_0|^2(\delta\Omega)^2}{\hbar^2\Omega^2}T\Omega\tilde{G}(\boldsymbol{k}'-\boldsymbol{k},\omega_{\boldsymbol{k}'}-\omega_{\boldsymbol{k}}), \tag{6.59}$$

where $\tilde{G}(\boldsymbol{k},\omega)$ is the Fourier transform of $G_\theta(\boldsymbol{r},t)$. In order to obtain a theoretical formula useful for comparison with the experiments, we have to sum this quantity over $\boldsymbol{k}'$ near $\boldsymbol{k}$, that is, over $\boldsymbol{K} = \boldsymbol{k}'-\boldsymbol{k}$ near 0, with weight $w(\boldsymbol{K})$ given by the $\boldsymbol{k}$ and $\boldsymbol{k}'$ distributions. We finally obtain the following first-order estimate for the decoherence parameter

$$\begin{aligned}
\epsilon \simeq \sum_{\boldsymbol{k}'\text{ near }\boldsymbol{k}} \overline{|\Delta_1|^2} &\Rightarrow \frac{\Omega}{(2\pi)^3}\int d^3\boldsymbol{K}\, w(\boldsymbol{K})\overline{|\Delta_1|^2} \\
&= \frac{|V_0|^2(\delta\Omega)^2}{\hbar^2}T\langle\rho\rangle\int\frac{d^3\boldsymbol{K}}{(2\pi)^3}\, w(\boldsymbol{K})\tilde{G}(\boldsymbol{K},\boldsymbol{K}\cdot\boldsymbol{v}_0) \\
&= \frac{1}{4}(\ln t_0)^2\left[1+\left(\frac{V_{0R}}{V_{0I}}\right)^2\right]f, \tag{6.60}
\end{aligned}$$

where we put $\omega_{\boldsymbol{k}'}-\omega_{\boldsymbol{k}} \simeq \boldsymbol{K}\cdot\boldsymbol{v}_0$, $\boldsymbol{v}_0$ being the particle velocity. The quantity f is a dimensionless constant given by

$$f = \frac{1}{\langle\rho\rangle T}\int\frac{d^3\boldsymbol{K}}{(2\pi)^3}\, w(\boldsymbol{k})\tilde{G}(\boldsymbol{K},\boldsymbol{K}\cdot\boldsymbol{v}_0). \tag{6.61}$$

Notice that f is written in terms of the spectral distribution function $\tilde{G}$.

We should remark that the value obtained perturbatively in Eq. (6.60) is sensible within the limit of validity of the expansion (6.45). On the other hand, the quantity $\overline{|\Delta_1|^2}$ becomes very large at very low values of the transmission probability t, i.e. at high values of $(\ln t)^2$ (in a first approximation t_0 can be set equal to t). Since the experimental data[145] seem to indicate that the deviation from the standard expression for the visibility is rather substantial, we have developed a nonperturbative approach to the calculation of ϵ in the previous subsection.

Chapter 7

NUMERICAL SIMULATIONS OF MEASUREMENT
PROCESSES

We have understood that the wave function collapse (WFC) by measurement
is just a physical process between an object quantum particle Q and a measuring (macroscopic) system D, giving rise to decoherence (or dephasing) on
the former. The process has been formulated within the framework of the
many-Hilbert-space (MHS) theory in Chapter 4, where we have introduced
the *decoherence parameter* ϵ in order to give a quantitative measure of the
(loss of) coherence. Even though some solvable models of the quantum measurement problem give us important insights and hints about this issue, as
has been shown in Chapter 5, the physical process taking place inside D when
particle Q interacts with the elementary constituents of D deserves further
clarification. Due to the complexity of the interaction and the huge number
($\sim 10^{23}$) of degrees of freedom involved, it would be desirable if we could solve
the physical process, on the basis of an appropriately chosen model, to understand how the coherence of Q is actually lost in practical cases. This can be
done *numerically*, at least for simple (yet realistic enough) models, without
any approximation.

In this chapter, we shall investigate the decoherence process from a numerical point of view, by studying the interaction between an "object particle" Q and a macroscopic object D made up of a huge number of elementary
constituents. The whole dynamics is obviously governed by the Schrödinger
equation. As a first trial, we shall neglect any inner structure of the elementary
constituents of D and assume that their interaction with Q is represented by
short-range potentials. These simplifications are acceptable if we are interested
only in the dephasing process on Q, and not in the change of the structure of
D which might lead to a signal production. (Remember that the latter process plays no essential role in the WFC, as has repeatedly been stressed in
this book.) More specifically, every interacting constituent (or every bunch of
constituents) will be schematized as a δ-potential, and the whole apparatus D
as a one-dimensional array of δ-potentials (a Dirac comb). This is nothing but

the Kronig-Penney model.[122,104,102,108,171]

Previous numerical simulations based on the same system[142] have shown that this simple model is able to reproduce correctly many different physical devices, like a "phase shifter" (that preserves the quantum coherence) or a "dephaser" (that provokes complete dephasing and works as a detector) or an "absorber" (that yields nonvanishing absorption), depending on the parameters involved. It is noteworthy that, against a widespread belief, absorption effects are not necessarily significant for the collapse of the wave function, because the loss of coherence (of which the decoherence parameter ϵ is an estimate) stems only from the fluctuations inherent in the interaction with a macroscopic apparatus. In the following, unlike in other numerical simulations of the measurement process by means of a Dirac comb,[127,142,144] we shall focus our attention on the role played by some particular combinations of the numerical constants characterizing the array of potentials.[107,129]

Our purpose is essentially twofold. First, we shall focus on the role played by the apparatus as a "dephaser." This is the reason why we shall not look at the mechanisms that generate signals in the apparatus and is somewhat implicit in our description of the apparatus in terms of a Dirac comb: The elementary constituents (represented by δ-potentials) are structureless and their scope is only to provoke dephasing on the object particle via a potential scattering. Second, the dephasing effects will be quantitatively analyzed by means of the decoherence parameter ϵ. We shall see that ϵ effectively plays the role of an "order parameter" for the wave function collapse, in the sense that it gives "critical" values in order that the quantum coherence be lost.

1   Dirac comb

Consider a one-dimensional array of N δ-potentials (Dirac comb)

$$V(x) = \sum_{\ell=1}^{N} \Lambda \, \delta(x - b_\ell), \qquad (7.1)$$

where Λ stands for the strength of the "elementary" interactions and b_ℓ for their locations. The transmission and reflection coefficients, T and R, of the whole barrier satisfy[107,129]

$$\begin{pmatrix} T \\ 0 \end{pmatrix} = e^{-ikb_N} Z \prod_{\ell=1}^{N-1} \left[e^{ikd_\ell \tau_3} Z \right] \begin{pmatrix} 1 \\ R \end{pmatrix}, \qquad (7.2)$$

where τ_3 is the third Pauli matrix, k the wave number of the Q particle and $d_\ell = b_{\ell+1} - b_\ell$. The transfer matrix Z has the explicit form

$$Z = \begin{pmatrix} 1 - i\zeta & -i\zeta \\ i\zeta & 1 + i\zeta \end{pmatrix}, \qquad (7.3)$$

where $\zeta = \Lambda/\hbar v$, v being the speed of Q. We shall set $b_1 = 0$.

So far, the internal motions of the elementary constituents of the apparatus have not been taken into account. These internal motions will give rise to
an intrinsic stochasticity of the parameters characterizing the configuration of
the apparatus. In our Dirac-comb model, this stochasticity will be modelled
as follows: since the interactions between the apparatus' constituents and the
incoming particles will take place in different parts of the apparatus, the positions b_ℓ, and therefore the relative spacings $d_\ell = b_{\ell+1} - b_\ell$, and also the total
number N of interactions will be subject to statistical fluctuations.

Equation (7.2) will be solved numerically for different values of d_ℓ and N,
which characterize the configuration of the potentials that each incoming particle meets: in the simulation, the average transmission/reflection coefficient
and probability will always be computed for $N_p = 1000$ events and the ensem-
ble average $\langle \cdots \rangle$ will be taken over a Gaussian distribution of d_ℓ and N, under
the assumption of ergodicity (see the discussion in Section 2.1 of Chapter 4).

We carefully avoid the case of resonance reflection arising from the lattice-
like structure of the Dirac comb, which may lead to a reflection probability
of order unity. This occurs when $k\langle d \rangle$ ($\langle d \rangle$ being the average spacing between
scatterers) is close to an integer multiple of π. In the simulation, in order to
reduce unwanted spurious effects, we shall always set

$$k\langle d \rangle = 4.5\pi. \tag{7.4}$$

We shall focus our attention on thermal neutrons interacting with atoms, so
that

$$\lambda = \frac{2\pi}{k} = 2\,\text{Å}, \qquad \langle d \rangle = 4.5\,\text{Å}. \tag{7.5}$$

The Gaussian distributions of the spacings and of the number of δ-potentials
are characterized by averages $\langle d \rangle$ and $\langle N \rangle$ and (normalized) standard deviations

$$\Delta d \equiv \delta d/\langle d \rangle, \qquad \delta d \equiv \sqrt{\langle (d - \langle d \rangle)^2 \rangle}, \tag{7.6}$$

$$\Delta N \equiv \delta N/\sqrt{\langle N \rangle}, \qquad \delta N \equiv \sqrt{\langle (N - \langle N \rangle)^2 \rangle}. \tag{7.7}$$

The values of these two parameters are varied in the ranges

$$0 \leq \Delta d \leq 0.5, \qquad 0 \leq \Delta N \leq 1. \tag{7.8}$$

The large fluctuations in the spacings (i.e. the large values of Δd) do not
seem appropriate to describe a rigid lattice. Indeed, there are situations in
which coherence is lost for $\Delta d > 0.1$. In such a case, the apparatus cannot be
viewed as a solid: A rigid lattice cannot work properly as a "detector," in the
sense that it does not provoke decoherence on the incoming particle. Notice

also that many detectors make use of thermodynamically unstable states, like gases or liquids in critical conditions, whose physical states are characterized by very large statistical fluctuations. The quantity ζ plays the role of a coupling constant, that can neither be too small, nor too big, in order to ensure the occurrence of some dephasing. In our simulation $\langle N \rangle$ will reach the value 10^4 and it has been found numerically that if $\zeta \ll 0.1$ no decoherence takes place. We therefore set in the following numerical study

$$\zeta = 0.1. \tag{7.9}$$

The values of the transmission probability $t \equiv \langle |T|^2 \rangle$ and decoherence parameter ϵ (or, equivalently, the transmission coefficient) depend on all the physical quantities we have considered so far. Since the numerical values of $\langle d \rangle$ and ζ are *fixed* according to Eqs. (7.5) and (7.9), respectively, we can state that, in general, $\Delta d, \Delta N$ and $\langle N \rangle$ determine t and ϵ. Our task is to clarify how the wave function collapse, i.e. the loss of coherence takes place ($\epsilon \to 1$) when the numerical values of $\Delta d, \Delta N$ and $\langle N \rangle$ change. For the sake of simplicity and clarity, we shall display only some of the main results of the simulations in the following way: First, the decoherence parameter ϵ and transmission probability t will be plotted as functions of $\langle N \rangle$ and ΔN, for several values of Δd. Second, they will be evaluated, for several values of ΔN, as functions of $\langle N \rangle$ and Δd. Third and last, their dependence on certain combinations of the parameters will be shown for several values of $\langle N \rangle$. We can derive approximate (analytic) expressions for ϵ and t under some plausible assumptions, which might give useful insight in establishing a so-called "design theory" of a detector (dephaser) in the future.

1.1 ϵ and t as functions of $\langle N \rangle$ and ΔN

Figure 7.1 displays the dependence of ϵ on $\langle N \rangle$ and ΔN, for several fixed values of Δd; Fig. 7.2 the behaviour of t as a function of the same parameters, and Fig. 7.3 the behaviour of ϵ versus $\langle N \rangle (\Delta N)^2$ for $\Delta d = 0$.

Observe that if $\Delta d = 0$ the transition region from a coherent behaviour ($\epsilon = 0$) to a totally decoherent one ($\epsilon = 1$) is very sharp and occurs along the lines $\langle N \rangle (\Delta N)^2 = $ const. This behaviour can be explained by noting that, for small ζ, the transmission coefficient T for a single δ-potential reads

$$T = \frac{1}{1 + i\zeta} = \frac{1}{\sqrt{1 + \zeta^2}} e^{-i \tan^{-1} \zeta} \simeq e^{-\frac{1}{2}\zeta^2} e^{-i\zeta}, \tag{7.10}$$

and by taking the average over N with a Gaussian distribution and neglecting small quantities, we obtain

$$t = \langle |T|^2 \rangle \simeq 1, \qquad |\langle T \rangle|^2 \simeq e^{-\langle N \rangle (\Delta N)^2 \zeta^2}, \tag{7.11}$$

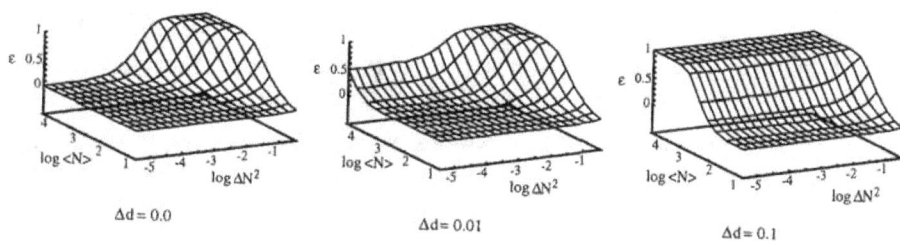

Fig. 7.1. Dependence of ϵ on $\langle N \rangle$ and ΔN, for some values of Δd. Observe the sharp transition region from $\epsilon = 0$ (coherence) to $\epsilon = 1$ (decoherence) when $\Delta d = 0$. See text.

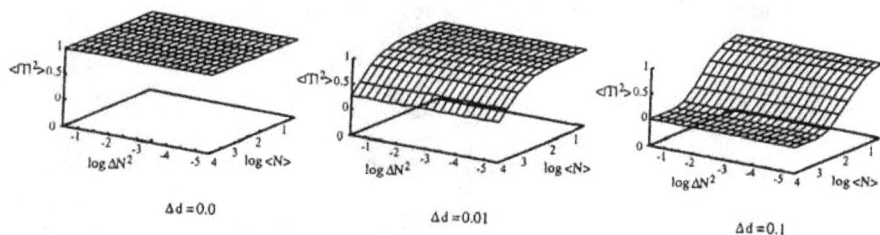

Fig. 7.2. Transmission probability $t = \langle |T|^2 \rangle$ vs $\langle N \rangle$ and ΔN.

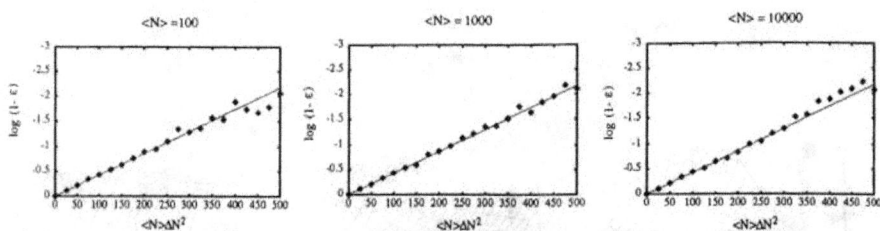

Fig. 7.3. Dependence of ϵ on $\langle N \rangle (\Delta N)^2$ for $\Delta d = 0$. The dotted line is due to the estimate Eq. (7.12). The diamond points are the numerical results. See text.

so that

$$\epsilon \simeq 1 - e^{-\langle N \rangle (\Delta N)^2 \zeta^2}. \tag{7.12}$$

In Fig. 7.3, the "diamond" points are the numerical results and the dotted lines are our approximate formula (7.12). The agreement is excellent.

The above reasoning explains the existence of a sharp transition along the lines $\langle N \rangle (\Delta N)^2 = $ const. when $\Delta d = 0$. This situation is physically interesting because it corresponds to the case of a neutron beam interacting with a macroscopic solid object, like a crystal: The spacing between different atoms of the macroscopic object is (almost) constant, but different neutrons impinge on different parts of the crystal, at slightly different angles, interacting therefore with a different number of elementary scatterers.

When the condition $\Delta d = 0$ (fixed spacings between adjacent potentials) does not hold anymore, additional randomization effects appear and the above approximations break down. This is shown in the other graphs of Figs. 7.1 and 7.2.

1.2 ϵ and t as functions of $\langle N \rangle$ and Δd

Figure 7.4 shows the dependence of ϵ on $\langle N \rangle$ and Δd, for several values of ΔN. The behaviour of $t = \langle |T|^2 \rangle$ as a function of the same parameters is shown in Fig. 7.5. For large values of Δd, this situation corresponds to a neutron beam interacting with a material made up of a liquid or a dilute solution or a gas. (A solid lattice would "melt" for such large values of Δd.)

It is interesting to notice that there is a "saturation" effect, in this case: If $\Delta d > 0.1$, the values of ϵ and t do not depend significantly on $\langle N \rangle$. The easiest way to understand this effect is to take the ensemble average of both sides of Eq. (7.2): Since the fluctuations of d_ℓ are independent for different ℓ,

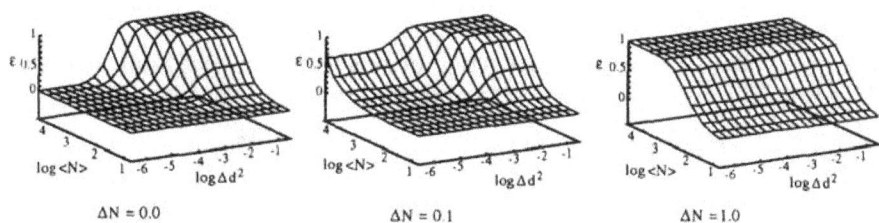

Fig. 7.4. Dependence of ϵ on $\langle N \rangle$ and Δd. Observe the "saturation" effect when $\Delta d > 0.1$. See text.

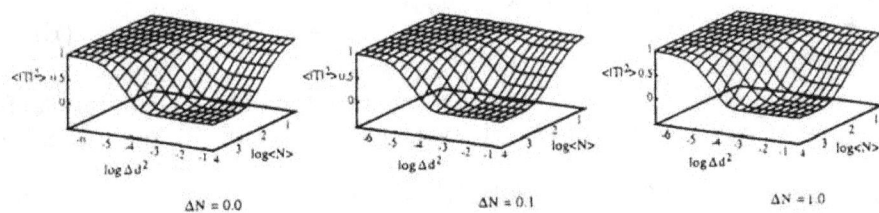

Fig. 7.5. t vs $\langle N \rangle$ and Δd.

$\langle T \rangle$ contains factors of the type

$$\langle e^{ikd_\ell} \rangle \simeq e^{ik\langle d \rangle} e^{-k^2\langle d \rangle^2 (\Delta d)^2/2}, \tag{7.13}$$

that vanish (washing away the dependence on Δd) when $\Delta d \gg 2/k\langle d \rangle \simeq 0.1$ [see Eq. (7.4)]. A similar situation occurs for t. Observe that this phenomenon takes place independently of the value of ΔN.

A more detailed explanation of this effect is given in Ref. 107, 129. When $\Delta N = 0$ (thus $\langle N \rangle = N$), we expand T in powers of the reflection coefficient $\mathcal{R} = -i\zeta/(1 + i\zeta)$ for a single δ-potential scattering:

$$T = \mathcal{T}^N \{1 + \alpha(\mathcal{R})\}$$
$$= \mathcal{T}^N \left\{ 1 + \mathcal{R}^2 \sum_{n=0}^{N-2} \mathcal{T}^{2n} \sum_{\ell=1}^{N-1-n} e^{2ik(d_\ell + \cdots + d_{\ell+n})} + O(\mathcal{R}^4) \right\}. \tag{7.14}$$

Since for large fluctuations of the spacings (i.e. for large Δd), $\langle e^{ikd_\ell} \rangle \simeq 0$, as can be seen in (7.13), we obtain, to second order in $\mathcal{R}$,

$$|\langle T \rangle|^2 \overset{\Delta d \to \text{large}}{\longrightarrow} \exp\left(-N|\mathcal{R}|^2\right), \tag{7.15}$$

$$t = \langle |T|^2 \rangle \overset{\Delta d \to \text{large}}{\longrightarrow} \exp\left(-N|\mathcal{R}|^2\right) \exp\left(|\mathcal{R}|^4 S_N^{(4)}\right), \tag{7.16}$$

$$\epsilon \overset{\Delta d \to \text{large}}{\longrightarrow} 1 - \exp\left(-|\mathcal{R}|^4 S_N^{(4)}\right). \tag{7.17}$$

Here we have used the relations $\mathcal{R}^2 \sim -|\mathcal{R}|^2$, $|\mathcal{T}|^2 = 1 - |\mathcal{R}|^2$ and

$$S_N^{(4)} \equiv \left\langle \left| \sum_{n=0}^{N-2} \mathcal{T}^{2n} \sum_{\ell=1}^{N-1-n} e^{2ik(d_\ell + \cdots + d_{\ell+n})} \right|^2 \right\rangle$$
$$\overset{\Delta d \to \text{large}}{\longrightarrow} \frac{|\mathcal{T}|^{4N} - N|\mathcal{T}|^4 + N - 1}{(|\mathcal{T}|^4 - 1)^2}. \tag{7.18}$$

Notice that these results are essentially due to a complete randomization of the phases acquired by multiple reflections, which leads to

$$\langle \alpha(\mathcal{R}) \rangle \overset{\Delta d \to \text{large}}{\longrightarrow} 0. \tag{7.19}$$

The agreement of the "perturbative" formulae (7.15)-(7.17) with the numerical results is excellent, as can be seen in Fig. 7.6.

On the other hand, for relatively small Δd and $\Delta N = 0$, no "saturation" effect is seen. In this case, one finds[107,129] that the average transmission coefficient is given by

$$\langle T \rangle \simeq \mathcal{T}^N \{1 - |\mathcal{R}|^2 S_N^{(2)}\} \tag{7.20}$$

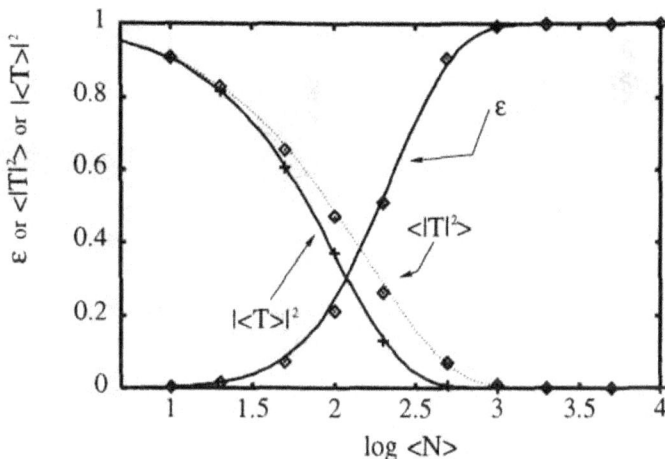

Fig. 7.6. "Saturation effect": Comparison of the "perturbative" formulae (7.15)-(7.17) with the numerical results. The curves are given by the formulae (7.15)-(7.17). The numerical results are represented by the points. ($\Delta N = 0$.)

to second order in $\mathcal{R}$. Here the quantity $\mathcal{S}_N^{(2)}$ is defined by

$$\mathcal{S}_N^{(2)} = \langle \sum_{n=0}^{N-2} \mathcal{T}^{2n} \sum_{\ell=1}^{N-1-n} e^{2ik(d_\ell + \cdots + d_{\ell+n})} \rangle$$

$$= \frac{s(s^N - Ns + N - 1)}{\mathcal{T}^2(s-1)^2}, \qquad s \equiv \mathcal{T}^2 \langle e^{2ikd} \rangle. \qquad (7.21)$$

Since the above expression (7.20), together with the relation $|\mathcal{T}|^2 = 1 - |\mathcal{R}|^2$, enables one to write

$$|\langle T \rangle|^2 \simeq \exp\left(-N_{\text{eff}}|\mathcal{R}|^2\right), \qquad (7.22)$$

with

$$N_{\text{eff}} \equiv N + 2\text{Re}\,\mathcal{S}_N^{(2)}, \qquad (7.23)$$

we are led to interpret N_{eff} as the number of potentials with which the Q particle *effectively* interacts. [Compare this with (7.15).] Under this interpretation, we may just follow a heuristic approach and replace N in (7.16) and (7.17) with N_{eff} to obtain

$$t = \langle |T|^2 \rangle \simeq \exp\left(-N_{\text{eff}}|\mathcal{R}|^2\right) \exp\left(|\mathcal{R}|^4 \mathcal{S}_{N_{\text{eff}}}^{(4)}\right), \qquad (7.24)$$

$$\epsilon \simeq 1 - \exp\left(-|\mathcal{R}|^4 \mathcal{S}_{N_{\text{eff}}}^{(4)}\right). \qquad (7.25)$$

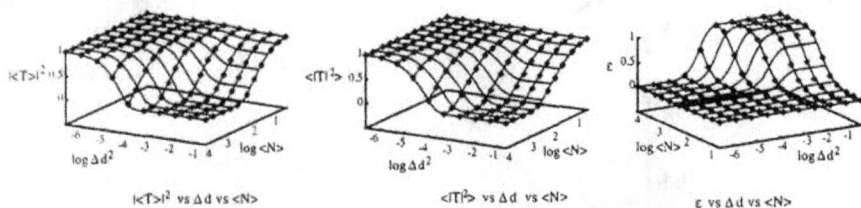

Fig. 7.7. General case: Comparison of the "perturbative" formulae (7.22), (7.24) and (7.25) with the numerical results. The solid curves are given by these formulae and the diamond points are the numerical results. ($\Delta N = 0$.)

Observe that the agreement of the "perturbative" formulae (7.22), (7.24) and (7.25) with the numerical results is excellent, as can be seen in Fig. 7.7.

The decoherence parameter ϵ is shown as a function of $N(\Delta d)^2$ when $\Delta N = 0$ in Fig. 7.8. It is interesting to notice that, for large N, ϵ becomes a function of the single variable $N(\Delta d)^2$. Like in the case discussed in Section 1.1, when $\Delta N \neq 0$ additional randomization effects appear. The resultant effect is seen in Figs. 7.4 and 7.5.

1.3   Analytic expressions for ϵ and t

Figure 7.9 shows how the decoherence parameter ϵ and transmission probability t depend on some particular combinations of the parameters for several fixed values of $\langle N \rangle$. In order to understand their behaviour and to derive appropriate analytic expressions for them, observe first that t turns out to be largely independent of ΔN. We are therefore led to assume that

$$t = \langle |T|^2 \rangle_{N,d} \simeq \langle |T|^2 \rangle_d$$
$$\simeq \exp\left(-N_{\text{eff}} |\mathcal{R}|^2\right) \exp\left(|\mathcal{R}|^4 \mathcal{S}_{N_{\text{eff}}}^{(4)}\right), \tag{7.26}$$

where we explicitly write which variables the averages are taken over. (See (7.24) for $\langle |T|^2 \rangle_d$.)

Moreover, if we assume[107,129]

$$\langle T \rangle_{N,d} = \langle e^{-iN\zeta} \rangle_N \langle T \rangle_d|, \tag{7.27}$$

we obtain

$$|\langle T \rangle_{N,d}|^2 \simeq |\langle e^{-iN\zeta} \rangle_N|^2 |\langle T \rangle_d|^2$$
$$\simeq \exp\left(-\zeta^2 \langle N \rangle (\Delta N)^2\right) \exp\left(-N_{\text{eff}} |\mathcal{R}|^2\right), \tag{7.28}$$

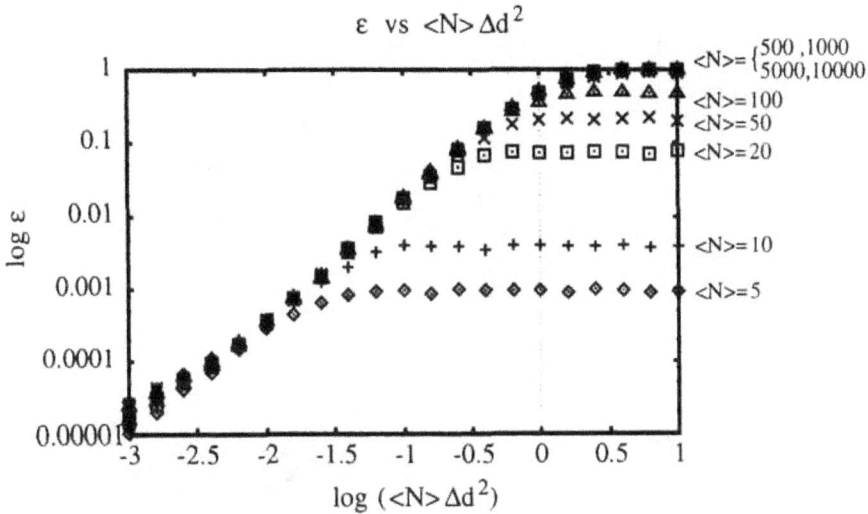

Fig. 7.8. Dependence of ϵ on $N(\Delta d)^2$ for several values of N. ($\Delta N = 0$.)

so that the decoherence parameter turns out to be

$$\epsilon \simeq 1 - \exp\left(-\zeta^2 \langle N \rangle (\Delta N)^2\right) \exp\left(-|\mathcal{R}|^4 \mathcal{S}^{(4)}_{N_{\text{eff}}}\right). \tag{7.29}$$

In spite of the crudeness of the above approximations, the agreement of these formulae with the numerical results is excellent, as can be seen in Fig. 7.9. Equations (7.26), (7.28) and (7.29) are our main results and cover all the particular cases hitherto considered.

1.4 Comments

It is interesting to observe that the parameters ΔN and Δd provoke decoherence in different ways: The value of the transmission probability t is practically ΔN-independent; on the other hand, it is strongly dependent on Δd. The different roles played by ΔN and Δd can be seen also in the phase diagrams for the transmission coefficient T_j ($j = 1, \ldots, N_p$), plotted in the complex plane. Interested readers should refer to Ref. 107. This observation is important for two independent reasons. First, this analysis sets the preliminary basis for a "design theory" of a quantum mechanical detector. The combined action of $\langle N \rangle$, ΔN and Δd, eventually leading to a complete loss of quantum mechanical coherence, can be suitably tailored and exploited in order to yield the desired type of dephasing (e.g. by transmitting the Q particle with high or low

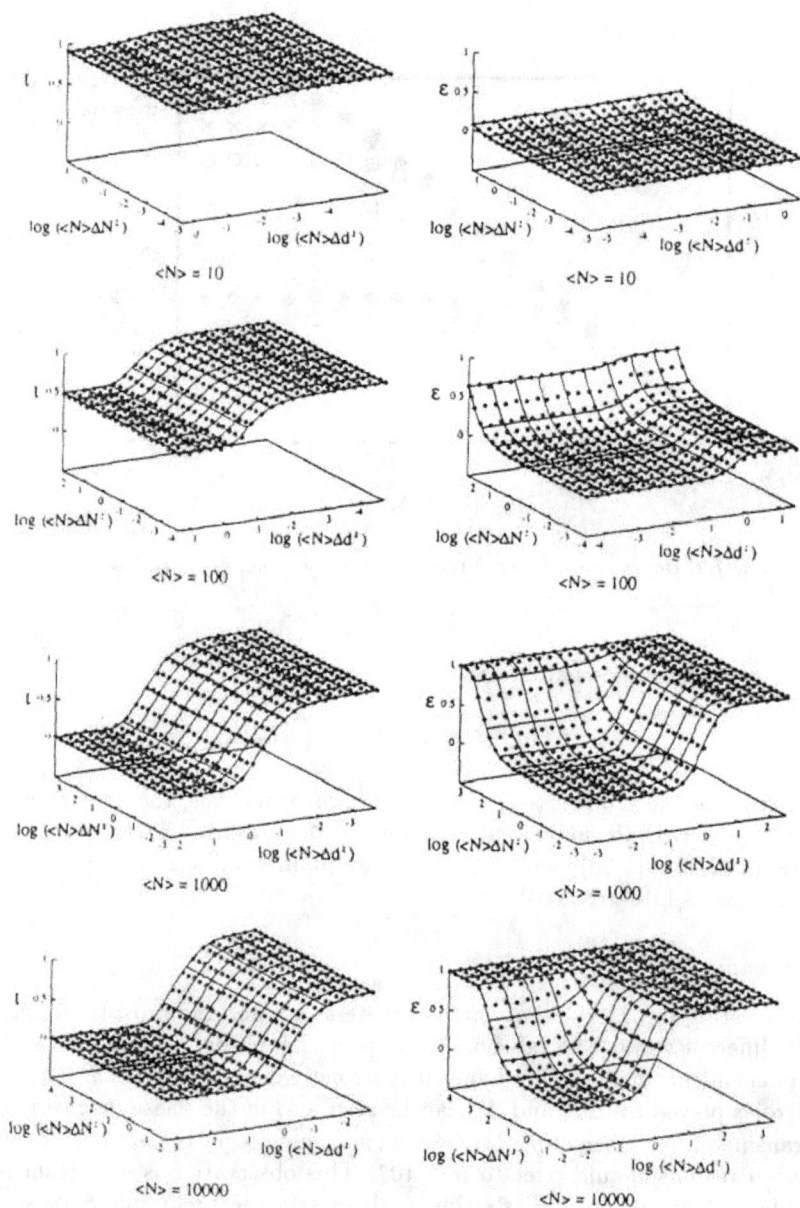

Fig. 7.9. Decoherence parameter ϵ and transmission probability t versus $\langle N \rangle (\Delta N)^2$ and $\langle N \rangle (\Delta d)^2$, for several values of $\langle N \rangle$. The solid curves are given by (7.26) and (7.29) and the diamond points are the numerical results.

probability). Second, this analysis can be important for the study of neutron interferometry experiments at very low transmission probability,[161,164,185] because it suggests that different physical devices, like a silicon crystal, a gaseous or a liquid absorber, can modify the coherence properties of an incoming neutron beam in many different ways. For example, one cannot expect a strong dependence on Δd for a crystal. On the other hand, an interesting dependence of ϵ on Δd can certainly be expected when a gaseous neutron absorber (say Ne) or a water solution of some highly absorbing material (say Gd) is inserted in an interferometer.[164] Incidentally, notice that a dependence on ΔN is to be expected both for a crystal or a gaseous device, because here geometrical factors play an important role.

An exhaustive discussion of these effects, beyond the qualitative estimates given here and in Refs. 145, 128, requires a different study and is not easy to perform, because it involves a careful estimate of the most important characteristics of the experimental setup. Such a program is at present under consideration.

However, it is possible to draw some general conclusions even from the results of our simulation: A collection of many different numerical simulations is shown in Fig. 7.10. In *all the cases investigated*, one observes that when $t = \langle |T|^2 \rangle$ is close to 0, in general $\epsilon \simeq 1$. In other words, it appears that in general, *decoherence effects cannot be avoided at low transmission probability*.

Notice that this phenomenon exists for all values of the parameters ΔN, Δd and $\langle N \rangle$. Even though the present numerical simulation neglects many important characteristics of a real experiment, it is difficult to believe that the behaviour displayed in Fig. 7.10 is just a coincidence. It is worth stressing that the statistical fluctuations of a real macroscopic apparatus can *never* be neglected, even in principle, because of finite-temperature effects and of the impossibility to isolate completely the apparatus from its environment.

2 Remarks

Our numerical simulation was based on a structureless detector, in order to shed light on its dephasing characteristics. As already emphasized, all the other functions of a "detector" have been neglected or suppressed in this analysis. It goes without saying that the detector model discussed in this chapter can be improved by introducing additional channels that take into account the change of the detector state (e.g. by introducing an internal structure for each δ-scatterer).

Equations (7.26), (7.28) and (7.29) summarize the main results of this analysis. The value of the transmission probability, transmission coefficient and decoherence parameter ϵ suffice to outline the essential features of the interaction between the Q particle and the apparatus.

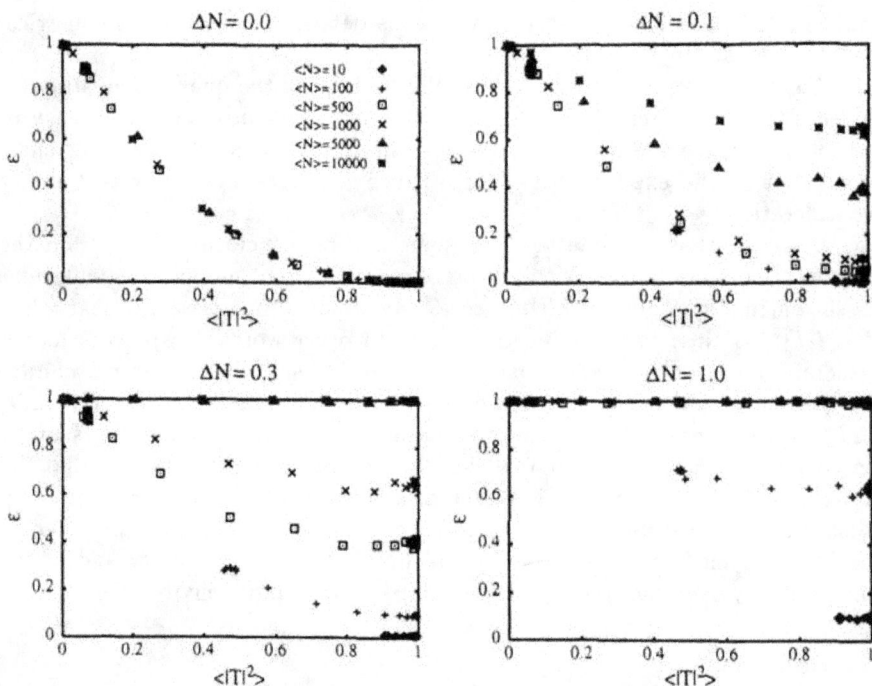

Fig. 7.10. "Summary" of different numerical simulations. The points refer to different values of ΔN, Δd and $\langle N \rangle$, i.e. $0 \leq \Delta N \leq 1$, $10 \leq \langle N \rangle \leq 10^4$ and $0 \leq \Delta d \leq 0.5$. In the first figure, $\Delta N = 0$, and all points lie on the same curve. In all cases, $\epsilon \to 1$ when $t \to 0$. *Dephasing and decoherence cannot be avoided at very low transmission probability.*

The dephasing effects have been quantitatively analyzed in terms of the decoherence parameter. We have seen that ϵ plays, in some sense, the role of an "order parameter" for the wave function collapse. This is simply an analogy to be considered with care. The idea that the quantum measurement process can be viewed as a sort of phase transition requires additional evidence and investigation. Nevertheless, Figs. 7.1, 7.4 and 7.9 are rather suggestive, in that they lead one to view the loss of quantum mechanical coherence as a phase transition of some sort. The idea that the "collapse" of the wave function can be viewed as a sort of phase transition was first proposed, as far as we know, by Ne'eman in a different context.[147]

Chapter 8

QUANTUM ZENO EFFECT

Let us discuss an interesting quantum mechanical issue which is very closely related to the measurement problem and is usually referred to as the quantum Zeno effect. This effect is a straightforward consequence of a very general property of the Schrödinger equation: If a quantum system evolves under the action of its Hamiltonian, in a small time δt its phase changes by $O(\delta t)$, while the square of its scalar product with the initial state changes by $O(\delta t^2)$. Therefore, by performing N measurements during a fixed time interval T the probability of observing the system in a state different from the initial one is approximately $\sim N\delta t^2 \propto N/N^2$, which becomes small as N is increased: In the limit of very frequent measurements, that state of the system does not change.

This effect was experimentally observed on an oscillating system.[43,84] Many physicists believed that this experiment provided clear evidence in support of the Copenhagen interpretation, in particular of what we called "naive WFC" in Chapter 1, consisting in the application of von Neumann's projection rule. However, the lively discussion[160,158,153,152] that followed the experiment showed that the experimental data can be explained *without* making use of projection operators: A purely dynamical analysis, fully based on the Schrödinger equation, yields the very same results. A recent overview on the quantum Zeno problem can be found in Ref. 81.

The purpose of this chapter is to discuss in depth the quantum Zeno effect and explain the above-mentioned issues. We shall start from a rather elementary introduction, then describe a few experimental tests and related ideas and finally comment on some subtle conceptual problems.

1 Historical introduction

1.1 Preliminaries

The temporal behaviour of quantum mechanical systems has been an interesting subject of investigation for many years and has recently attracted the attention of several researchers. This is due, in particular, to the curious features in the short-time region, which lead to seemingly paradoxical results.

The seminal work by Gamow[64] on the exponential decay law, as well as its derivation by Weisskopf and Wigner,[200] are based on the assumption that a pole near the real axis of the complex energy plane dominates the temporal evolution of the quantum system. This assumption leads to a spectrum of the Breit-Wigner type.[33] However, it was later understood that a purely exponential decay law can neither be expected for very short,[120,57] nor for very long times.[76,139,99] It soon became clear that the domain of validity of the exponential law is limited: the long-time power tails and the short-time quadratic behaviour are unavoidable consequences of the Paley-Wiener theorem[151] on Fourier transforms and of the boundedness of the energy spectrum, respectively. The temporal behaviour of quantum mechanical systems is reviewed in the Appendix of this book; A thorough discussion is given in Refs. 132, 92, 55, 58, 36, 81.

Moreover, the short-time quadratic behaviour can be exploited in order to increase the "survival probability" of an unstable quantum system.[21,100,59,46] This idea hinges upon the notion of quantum measurement, in the sense of what we called the "naive" Copenhagen interpretation in Chapter 1, namely the von Neumann projection postulate.

A few years later Misra and Sudarshan proved, by a rigorous mathematical theorem,[125] that the temporal evolution of a quantum system can be halted in the so-called limit of continuous observation and introduced the notion of "quantum Zeno paradox"—Zeno was the Eleatic philosopher famous for his paradoxical arguments against the philosophical notion of "becoming:" His sped arrows would never reach their target, if closely looked at. It is remarkable that von Neumann had realized[196] (already in 1932 !) that a quantum mechanical state can be "steered" into other states by a series of measurements in rapid succession. In particular, if the final state coincides with the initial one, the evolution of a quantum system can be halted (or at least dramatically slowed down).

Our interest in the quantum Zeno effect and the short-time behaviour of quantum-mechanical systems stems from the subtle role played by quantum measurements, in this context. There are several problems that deserve a thorough discussion: Has the notion of "continuous observation" a sensible physical interpretation, or is it rather an approximation? Is it really necessary to perform a series of quantum measurements in rapid succession, in order to obtain a quantum Zeno effect? We shall endeavour to tackle these questions in this chapter. We shall argue that the limit of continuous observation is unphysical and must be regarded as a mathematical idealization: it can be useful, but one should not push this concept too far. Moreover, we shall see that it is *not* necessary to make use of the notion of quantum measurement, in order to obtain Zeno-type dynamics. The first (to our knowledge) who shed light on this important point was Peres,[157] who realized that the quantum Zeno effect can be given a dynamical explanation, without making use of projection operators à la von Neumann. The analysis of the quantum Zeno effect in dynamical terms will also bring to light some fundamental limits of von Neumann's projection rule formalism and of the naive Copenhagen interpretation.

In this chapter we shall discuss some recent experimental and theoretical developments on the quantum Zeno effect (QZE). For review articles on these topics, see Refs. 58, 132, 81. A mathematically rigorous treatment can be found in Ref. 55.

1.2 Philosophical touch

Before analyzing the quantum Zeno effect, it is useful (and hopefully amusing) to summarize, from a physicist's point of view, the philosophy of Zeno and of his master Parmenides. Parmenides was the most prominent figure of the Eleatic school of philosophers. Elea was a small Italian town, geographically and culturally belonging to the Mediterranean region called "Magna Graecia." Parmenides was a profound thinker. He was influenced by the philosophy of Pythagoras and was exposed to the charm of mathematics. Russell remarks,[172] however, that at that time mathematics was entangled with mysticism: Science, as we know it today, was yet to be born.

What makes Parmenides important was his belief that senses are deceptive and our perception of reality in continuous change is a mere illusion. He believed in a unique, entire, indivisible *Truth* (his famous concept of "being"), and could not accept the idea that this Truth could undergo any change ("becoming") or even be composed of smaller entities. He was so convinced of this idea that would not hesitate to push it to its most extreme and paradoxical consequences.

Zeno was the most brilliant disciple of Parmenides. He was an unequaled orator, even though his contributions to the philosophy of "being" were not as important as his master's. It is very probable (but the historical data are uncertain) that Parmenides and Zeno visited Athens, about 450 B.C., to meet Socrates and discuss philosophy with him. The reader should realize that to undertake such a trip (from the South of Italy to the Eastern part of Greece), at those times, was a major event. Nowadays, such a philosophical discussion would probably take place by email and would undoubtedly be less effective.

The dialogue between Socrates and the two Eleatic philosophers is one of the most renowned debates in the history of Greek philosophy. Socrates won, but was profoundly influenced by Parmenides' ideas and by Zeno's unequaled qualities of orator. It is even sometimes said that Zeno is the inventor of dialectic, that was subsequently so widely used by Socrates.

Zeno liked to put forward many paradoxical examples, in order to support his master's credo. In this sense, he can be considered a forerunner of sophism. Two of his arguments in support of Parmenides' philosophical viewpoints are most famous. In the first one, Achilles cannot reach a turtle because when the former has arrived at the position previously occupied by the latter, the turtle has already moved away from it, and so on *ad infinitum*. The second of Zeno's paradoxes concerns us more directly: a sped arrow never reaches its target, because at every instant of time, by looking at the arrow, we clearly see that it occupies a unique and definite position in space. At every moment the arrow is therefore immobile, and by summing up so many "immobilities" it is clearly impossible, according to Zeno, to obtain motion. If he lived nowadays, Zeno would probably consider photographs (in which all moving objects are still) as the best proof in support of his ideas. It would be interesting to ask his opinion about movies.

It is clear that both paradoxes arise from the lack of the idea of *infinitesimal*. We know, nowadays, that modern infinitesimal calculus easily resolves the above-mentioned two paradoxes by introducing the concept of velocity as the derivative of position with respect to time: it takes a very small time to cover a very small distance, where "very small" times and distances are infinitesimal of the same order. Achilles will reach the turtle in a *finite* time and the arrow will move with a *nonvanishing* velocity. Nevertheless, one should not overlook the fact that Zeno basically aimed at giving very provocative arguments against the concept of "becoming," in order to support the philosophy of "being" and ridicule the critics of Parmenides.

It is interesting and somewhat amusing that some quantum mechanical states, under particular conditions, behave in a way that is reminiscent of Zeno's paradoxes.

2 Short-time behaviour and quantum Zeno effect

2.1 Classical vs quantum mechanical decays

Let us start by outlining the main differences between the classical and the quantum mechanical evolution laws. In classical physics one heuristically assumes that the decay probability per unit time is a constant Γ (the inverse of the "lifetime" τ_E), that is characteristic of the system investigated and does not depend on other quantities, such as the total number N of unstable systems, their past history and the environment surrounding them. Let the (very large) number of systems at time t be $N(t)$; the number of systems that will decay in the time interval dt is

$$-dN = N\Gamma dt = N dt/\tau_{\mathrm{E}}, \qquad (8.1)$$

which yields

$$N(t) = N_0 e^{-t/\tau_{\mathrm{E}}}, \qquad (8.2)$$

where $N_0 = N(0)$. The "survival" or "nondecay probability" reads

$$P_{\mathrm{cl}}(t) = \frac{N(t)}{N_0} = e^{-t/\tau_{\mathrm{E}}} \simeq 1 - t/\tau_{\mathrm{E}} + \cdots, \qquad (8.3)$$

where the expansion holds at short times ($t \ll \tau_{\mathrm{E}}$). The above derivation is usually found in elementary physics textbooks. However, the assumptions underpinning it are delicate, for they reflect the basic features of a Markoffian process, in which memory and/or collective effects are absent.

What about quantum mechanics? Let $|a\rangle$ be the wave function of a given quantum system at time $t = 0$. The evolution is governed by the unitary operator $U(t) = \exp(-iHt/\hbar)$, where H is the Hamiltonian[a] (see Section 2.2 of Chapter 2). The survival or nondecay probability at time t is defined as the square modulus of the survival amplitude

$$P(t) = |\langle a|e^{-iHt/\hbar}|a\rangle|^2. \qquad (8.4)$$

By assuming that $|a\rangle$ is normalizable and belongs to the domain of definition of H (so that all moments of H in the state $|a\rangle$ are finite), one gets the short-time expansion

$$P(t) = 1 - t^2/\tau_{\mathrm{Z}}^2 + \cdots, \qquad (8.5)$$

$$\tau_{\mathrm{Z}}^{-1} \equiv \Delta H = \left(\langle a|H^2|a\rangle - \langle a|H|a\rangle^2\right)^{1/2}, \qquad (8.6)$$

which is quadratic in t and therefore yields a vanishing decay rate for $t \to 0$. [It is assumed that ΔH is nonvanishing, or, in other words, that $|a\rangle$ is not an eigenstate of H. Otherwise one simply gets $P(t) = 1$.] The short-time quadratic behaviour is in manifest contradiction with the exponential law (8.3), that predicts an initial nonvanishing decay rate τ_{E}^{-1}. We shall refer to the quantity τ_{Z} as the "Zeno time," for reasons that will soon become apparent.

---

[a] Here and in the following, we shall suppress, for simplicity, the caret on operators.

2.2   Quantum Zeno effect and quantum Zeno paradox

The quantum mechanical vanishing decay rate at short times can be exploited in order to slow down the decay process. Suppose we perform N measurements at equal time intervals $\tau = T/N$, in order to ascertain whether the system is still in its initial state. After each measurement, the system is "projected" onto the quantum mechanical state representing the result of the measurement and the evolution starts anew with a vanishing decay rate, according to (8.5). The probability of observing the initial state at the final time $T = N\tau$, after having performed the N above-mentioned measurements, reads

$$P^{(N)}(T) = [P(\tau)]^N = [P(T/N)]^N \simeq \left(1 - \frac{1}{\tau_Z^2} \left(\frac{T}{N} \right)^2 \right)^N \overset{N \text{ large}}{\sim} e^{-T^2/\tau_Z^2 N}.$$
(8.7)

Notice that both T and N are finite, in the above. Equation (8.7) displays the quantum Zeno effect: Repeated observations slow down the evolution and increase the probability that the system is still in the initial state at time T.

The so-called limit of continuous observation is obtained by letting $N \to \infty$. In this case one obtains

$$P^{(N)}(T) \simeq \left(1 - \frac{1}{\tau_Z^2} \left(\frac{T}{N} \right)^2 \right)^N \overset{N \to \infty}{\longrightarrow} P(T) = 1.$$
(8.8)

Infinitely frequent observations halt the evolution and "freeze" the initial state of the quantum system. This is the quantum mechanical version of the Zeno paradox: Zeno's arrows, although sped, do not move and never reach their target, if carefully looked at; analogously, the quantum mechanical wave function, although "sped" under the action of its Hamiltonian, does not evolve, if it is continuously observed.

There are several reasons why the result (8.8) is to be viewed as para-. doxical. First of all, the $N \to \infty$ limit is practically unattainable, in a real experiment. More important, such a limit is unphysical, for it is in contradiction with Heisenberg's uncertainty principle,[66,131,193,155] as will be discussed in Section *3.3*. Finally, the above analysis assumes that a measurement process occurs instantaneously. This is a very misleading point: No physical process can occur instantaneously; therefore a quantum mechanical measurement—which, as we hope to have been clarifying throughout this book, is nothing but a concrete *physical* process—must take place during a certain finite elapse of time. As a consequence, one cannot assume that it is possible to perform an infinite number of measurements in the finite elapse of time T. This point will be discussed in Sections *3.2* and *3.3*.

It is then clear that it is much more reasonable to start one's considerations from formulae like (8.7), where both T and N are finite. On the other hand, it is well known that there are many physical situations in which the temporal behaviour of a quantum system can be considered approximately exponential, as in (8.3). It becomes therefore interesting to understand under which conditions the QZE can be observed.

In the above discussion, we have assumed that the Gaussian period is much shorter than the exponential one. This assumption, however, might turn out to be rather simplistic, because the possible resultant stochastic processes can be very different, depending on the relative magnitudes of τ_Z and τ_E, as discussed many years ago.[189,72]

Moreover, the present discussion is complicated by the presence of fixed T and N. Rewrite (8.7) as

$$P^{(N)}(T) \overset{N \text{ large}}{\sim} e^{-T\tau/\tau_Z^2}, \tag{8.9}$$

where $\tau = T/N$ is the time elapsed between successive measurements, and compare the above formula with $P_{cl}(T)$ in Eq. (8.3). Suppose you perform an experiment on an unstable quantum system, in order to check whether it has decayed or not. Then the quantum mechanical law can be found to be considerably different from exponential if

$$e^{-T/\tau_E} \ll e^{-T\tau/\tau_Z^2} \quad \Longleftrightarrow \quad \tau \ll \frac{\tau_Z^2}{\tau_E}. \tag{8.10}$$

This can be quite a serious constraint on τ, by virtue of the (expected) smallness of τ_Z and of the ratio τ_Z/τ_E. On the other hand, if

$$\tau \simeq \frac{\tau_Z^2}{\tau_E}, \tag{8.11}$$

then the exponential and the Gaussian regions are practically indistinguishable. The actual temporal behaviour to be observed depends therefore on the relative magnitude of the parameters τ_Z and τ_E and on the experimenter's ability to make τ (the time interval between successive measurements) small. This point was also discussed in Refs. 176 and 132.

2.3 Misra and Sudarshan's formulation

It is convenient to review the formulation of the quantum Zeno paradox as given by Misra and Sudarshan.[125] Even though other authors[21,100,59,46] had already brought to light the curious features of the short-time behaviour of quantum systems a few years before, this formulation blends mathematical rigor and physical insights into the Copenhagen interpretation. This enables one to pinpoint the limits inherent in such a description and to elucidate several conceptual issues.

In their seminal paper,[125] Misra and Sudarshan endeavoured to define "the probability $\mathcal{P}(T)$ that no decay is found *throughout the interval* $\Delta = [0, T]$ when the initial state of the system was known to be ρ_0." (Italics in the original. Some symbols have been changed. If the initial state is pure [as in (8.4)-(8.6)], then $\rho_0 = |a\rangle\langle a|$.)

There is no obvious and unique definition, in quantum mechanics, for such a probability $\mathcal{P}$ of "continuous observation throughout a time interval." The definition they proposed involves a limiting procedure:

$$\mathcal{P}(T) \equiv \lim_{N\to\infty} P^{(N)}(T), \tag{8.12}$$

where $P^{(N)}(T)$ is essentially the quantity defined in Eq. (8.7) and will be unambiguously defined below [see Eq. (8.19)].

The derivation of the quantum Zeno paradox is entirely based on what we called the naive Copenhagen interpretation in Chapter 1. Notice the fundamental role played by the projection operator $\mathcal{O}$ in the following and remember that its effect on the state of the system is supposed to take place *instantaneously*.

We prepare our system Q in the unstable state ρ_0 and let $\mathcal{O}$ be the projection operator over the subspace of the undecayed states. The evolution is governed by the unitary operator $U(t) = \exp(-iHt/\hbar)$, where H is a semi-bounded Hamiltonian. By definition,

$$\rho_0 = \mathcal{O}\rho_0\mathcal{O}, \qquad \mathrm{Tr}[\rho_0\mathcal{O}] = 1. \tag{8.13}$$

In the following, we shall neglect the (trivial) normalization factors of the density matrix. We perform a measurement at time τ, in order to check whether Q is still undecayed. According to the naive Copenhagen interpretation, if Q is found undecayed, its state instantaneously "collapses" into

$$U(\tau)\rho_0 U^\dagger(\tau) \to \mathcal{O}U(\tau)\rho_0 U^\dagger(\tau)\mathcal{O}, \tag{8.14}$$

so that the probability of finding the system undecayed is given by

$$P(\tau) = \mathrm{Tr}[U(\tau)\rho_0 U^\dagger(\tau)\mathcal{O}]. \tag{8.15}$$

We then repeat the same measurement at regular time intervals τ; each measurement is instantaneous. After N measurements, at time $T = N\tau$, the state of Q reads

$$\rho^{(N)}(T) = V_N(T)\rho_0 V_N^\dagger(T), \qquad V_N(T) \equiv [\mathcal{O}U(T/N)\mathcal{O}]^N, \tag{8.16}$$

where we made use of (8.13) and (8.14). The probability of finding the system undecayed is given by

$$P^{(N)}(T) = \mathrm{Tr}\left[V_N(T)\rho_0 V_N^\dagger(T)\right]. \tag{8.17}$$

In the so-called limit of continuous observation ($N \to \infty$), provided some technical requirements are satisfied and some limiting operators exist, the final state and the probability of observing the system still undecayed read respectively

$$\tilde{\rho}(T) \equiv \lim_{N \to \infty} \rho^{(N)}(T) = \mathcal{V}(T)\rho_0 \mathcal{V}^\dagger(T), \tag{8.18}$$

$$\mathcal{P}(T) \equiv \lim_{N \to \infty} P^{(N)}(T) = \text{Tr}\left[\mathcal{V}(T)\rho_0 \mathcal{V}^\dagger(T)\right], \tag{8.19}$$

where

$$\mathcal{V}(T) \equiv \text{s--}\lim_{N \to \infty} V_N(T), \qquad \mathcal{V}^\dagger(T)\mathcal{V}(T) = \mathcal{O}, \tag{8.20}$$

s denoting a strong limit in the operator topology. This implies that

$$\mathcal{P}(T) = \lim_{N \to \infty} P^{(N)}(T) = \text{Tr}\left[\rho_0 \mathcal{O}\right] = 1 : \tag{8.21}$$

If the particle is continuously observed, in order to check whether it has decayed or not, it is "frozen" in its initial (undecayed) state and is never found to decay! This is the "quantum Zeno paradox."

The reader must have noticed the fundamental role played by the assumption that measurements occur instantaneously. The limit of continuous observation ($N \to \infty$) is taken while keeping T finite and letting $\tau = O(1/N) \to 0$. Remember that τ is the time interval between successive measurements. The time actually spent in performing the measurement is not taken into account! This is obviously a mathematical approximation: No physical process (let alone quantum measurements) can occur instantaneously. This point will be reconsidered and carefully discussed in Section *3.3*, where the $N \to \infty$ limit will be shown to be in contradiction with the Heisenberg uncertainty principle, and in Section *3.2*, where we will show that one can obtain the same result by means of a purely dynamical process (that makes use of unitary evolutions and does not involve any projections, provided a final measurement is carried out).

As a matter of fact, the above derivation of quantum Zeno paradox is very instructive, in our opinion: One is really led to wonder about the physical meaning of the operator $\mathcal{O}$, and about the reason why a "projection," unlike all other physical phenomena, should occur instantaneously. These problems are thoroughly discussed in this book.

3   Experimental tests

3.1   Experiment with three-level ion

In the last few years much renewed interest in the QZE was triggered by an interesting idea put forward by Cook[43] and the subsequent experiment performed by Itano, Heinzen, Bollinger and Wineland.[84] The meaning of this experiment has been hotly debated: An (incomplete!) list of references is Refs. 160, 83, 158, 14, 85, 61, 80, 153, 148, 22, 152, 4, 20, 15. It is now widely believed that the experimental results can be explained by making use of unitary dynamics.[160,152]

However, it took some time to realize[133] that the experiment is in fact at variance with Misra and Sudarshan's original definition (8.12). Indeed, both Cook's interesting proposal and Itano *et al.*'s beautiful experiment investigate the probability of finding the initial state *at* time t, *regardless* of the state of the system throughout the time interval $[0, t]$. Let us discuss this point in detail.

The experiment is performed on a three-level atomic system (a $^9Be^+$ ion) in an rf cavity. The ion energy level configuration is $E_1 < E_2 < E_3$ and an rf field of frequency $\omega = (E_2 - E_1)/\hbar$ provokes Rabi oscillations between levels #1 and #2. In the rotating wave approximation and in absence of detuning, the Bloch vector $\boldsymbol{R} \equiv (R_1, R_2, R_3)$ undergoes a precession around $\boldsymbol{\omega} \equiv (\omega, 0, 0)$ according to the equation

$$\dot{\boldsymbol{R}} = \boldsymbol{\omega} \times \boldsymbol{R}. \tag{8.22}$$

As is well known, the Bloch vector is expressed in terms of the density matrix ρ according to

$$\begin{aligned} R_1 &\equiv \rho_{21} + \rho_{12}, \\ R_2 &\equiv i(\rho_{12} - \rho_{21}), \\ R_3 &\equiv \rho_{22} - \rho_{11} \equiv P_2 - P_1, \end{aligned} \tag{8.23}$$

where $P_j \equiv \rho_{jj}$ is the probability that the atom is in level #j ($j = 1, 2$). Observe that the third component of the Bloch vector simply expresses the difference between the populations of levels #1 and #2, while its first two components are expressed in terms of the off-diagonal elements of the density matrix. Notice also that, since $P_1 + P_2 = 1$,

$$P_2 = \frac{1}{2}(1 + R_3), \tag{8.24}$$

$$P_1 = \frac{1}{2}(1 - R_3). \tag{8.25}$$

The solution of Eq. (8.22), with initial condition $\boldsymbol{R}(0) \equiv (0, 0, -1)$ [i.e. $P_1(0) = 1, P_2(0) = 0$] reads

$$\boldsymbol{R}(t) = (0, \sin \omega t, -\cos \omega t) \tag{8.26}$$

and if the transition between the two levels is driven by an on-resonant π pulse, of duration $T = \pi/\omega$, only level #2 is finally populated:

$$P_2(T) = 1. \tag{8.27}$$

On the other hand, assume that you perform N measurements at equal time intervals $\tau = \pi/N\omega$ by provoking transitions from level #1 to level #3, followed by rapid spontaneous emissions of photons. This can be accomplished by irradiating the atom with very short optical pulses of frequency $\omega_0 = (E_3 - E_1)/\hbar$, and by choosing the level configuration in such a way that the spontaneous decay $3 \to 1$ is strongly favored, while the decay $3 \to 2$ is forbidden. In this way, the atom is inferred to be in level #1 if a spontaneously emitted photon of frequency ω_0 is observed, while it is in level #2 if no spontaneously emitted ω_0-photon is observed (the latter is nothing but a negative-result measurement, as discussed in Section *1.3* of Chapter 1). It is instructive to discuss what happens in terms of the naive Copenhagen interpretation: Every ω_0 pulse "projects" the atom into level #1 or #2 and suddenly makes $R_{1,2}$ vanish (remember that $R_{1,2}$ depend on the off-diagonal elements of the density matrix), while it leaves R_3 (that depends on the populations) unaltered. It is easy to see that after the first measurements, at time $\tau = \pi/N\omega$,

$$R(\tau) = [0, 0, -\cos(\pi/N)] \equiv R^{(1)}, \qquad (8.28)$$

and after N measurements, at time $T = N\tau = \pi/\omega$,

$$R(T) = [0, 0, -\cos^N(\pi/N)] \equiv R^{(N)}, \qquad (8.29)$$

so that the probabilities that the atom is in levels #2 and #1 at time T, after the N measurements, read

$$P_2^{(N)}(T) = \frac{1}{2}\left[1 + R_3^{(N)}\right] = \frac{1}{2}\left[1 - \cos^N(\pi/N)\right], \qquad (8.30)$$

$$P_1^{(N)}(T) = \frac{1}{2}\left[1 - R_3^{(N)}\right] = \frac{1}{2}\left[1 + \cos^N(\pi/N)\right], \qquad (8.31)$$

respectively. Since $P_2^{(N)}(T) \to 0$ and $P_1^{(N)}(T) \to 1$, as $N \to \infty$, this is interpreted as the quantum Zeno effect. The experimental results[84] are in very good agreement with the above formulae. However, this is *not* the quantum Zeno effect, according to the definition (8.12): Equation (8.30) [(8.31)] expresses only the occupation probability of level #2 [#1], namely the probability that the atom is in level #2 [#1] at time T, after N measurements, *independently* of its past history. As a matter of fact, repopulation effects of level #2 from level #1 and *vice versa* are not prevented. In order to understand this (rather subtle) point, let us look explicitly at the first two measurements.

After the first measurement, Eq. (8.28) yields

$$R_3^{(1)} = -\cos\frac{\pi}{N} = P_2^{(1)} - P_1^{(1)}, \qquad (8.32)$$

$$P_1^{(1)} \equiv \cos^2\frac{\pi}{2N}, \qquad P_2^{(1)} \equiv \sin^2\frac{\pi}{2N}, \qquad (8.33)$$

where $P_j^{(1)}$ is the occupation probability of level #j ($j = 1, 2$) at time $\tau = \pi/N\omega$, after the first ω_0 pulse. After the second measurement, from Eq. (8.29) one obtains

$$R_3^{(2)} = -\cos^2 \frac{\pi}{N} = P_2^{(2)} - P_1^{(2)}, \tag{8.34}$$

$$P_1^{(2)} \equiv \cos^4 \frac{\pi}{2N} + \sin^4 \frac{\pi}{2N}, \quad P_2^{(2)} \equiv 2\sin^2 \frac{\pi}{2N} \cos^2 \frac{\pi}{2N}, \tag{8.35}$$

where $P_j^{(2)}$ ($j = 1, 2$) are the occupation probabilities at time $2\tau = 2\pi/N\omega$. Look at Fig. 8.1: $P_1^{(2)}$ is *not the survival probability* of level #1: It is the probability that level #1 is populated at time $t = 2\pi/N$, including the possibility that the transition $1 \to 2 \to 1$ took place, with probability $\sin^2 \frac{\pi}{2N} \cdot \sin^2 \frac{\pi}{2N} = \sin^4 \frac{\pi}{2N}$. The survival probability, according to (8.12), is

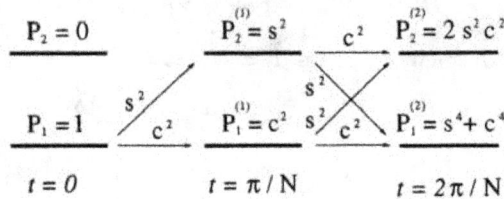

Fig. 8.1. Transition probabilities after the first two measurements ($s \equiv \sin \frac{\pi}{2N}$ and $c \equiv \cos \frac{\pi}{2N}$).

the probability that the atom is found in level #1 *both* in the first and second measurements, and is given by $\cos^4 \frac{\pi}{2N}$.

The repopulation effect that takes place after N measurements is described in Ref. 133 and brings to light the presence of a binomial distribution in the probabilities (8.30)-(8.31): indeed it is not difficult to realize that

$$P_2^{(N)}(T) = 1 - \sum_{n \text{ even}} \binom{N}{n} \sin^{2n} \frac{\pi}{2N} \cos^{2(N-n)} \frac{\pi}{2N}, \tag{8.36}$$

$$P_1^{(N)}(T) = \sum_{n \text{ even}} \binom{N}{n} \sin^{2n} \frac{\pi}{2N} \cos^{2(N-n)} \frac{\pi}{2N}, \tag{8.37}$$

where $\sum_{n \text{ even}}$ is the sum over all even values of n between 0 and N. This clearly shows that Eqs. (8.30)-(8.31) include all possible transitions between levels #1 and #2, in such a way that at time T the system is, say, in level #2 after having made an odd number ($n = 1, 3, \ldots$, etc.) of transitions between levels #1 and #2. The result (8.31) is conceptually very different from Misra and Sudarshan's survival probability (8.12). The correct formula for the survival probability, in the present case, is obtained by considering *only* the $n = 0$ term in (8.37):

$$\mathcal{P}_1^{(N)}(T) = \cos^{2N} \frac{\pi}{2N}. \tag{8.38}$$

This is the genuine "survival" probability, namely the probability that level #1 is populated at every measurement, at times $n\tau = nT/N$ $(n = 1, \ldots, N)$. This formula was first given by von Neumann.[195]

It must be emphasized that we are not criticizing the soundness of the experiment by Itano *et al.* Indeed, their experimental results are in excellent agreement with Eq. (8.30). We only claim that this experiment, although correctly performed, is conceptually at variance with the original idea on the QZE, as defined by Misra and Sudarshan.

It should be stressed that the above conclusions hold true for all those situations in which the temporal behaviour of the system under investigation is of the oscillatory type, and no precautions are taken in order to prevent repopulation of the initial state. The subtle point, in relation with Cook's proposal and Itano *et al.*'s experiment, is that the state of the atom is *not* observed at intermediate times.[83,133] As a matter of fact, its observation would probably raise difficult technical problems, for one should be able to judge and select, after each measurement pulse, *which* atoms are in level #1 and discard those atoms that are in level #2.

3.2 Experiment with neutron spin

Let us look at another experimental proposal to test the QZE, that makes use of neutron spin.[153] This experiment does not suffer from the "repopulation" drawback described in Section *3.1*. The interest of this proposal lies in the idea that it is possible to obtain both QZE (8.7) or (8.17) and quantum Zeno paradox (8.8) or (8.21) by making use of a purely dynamical process, that does not involve projection operators, as in Section *2.3*.

Consider the experimental setup sketched in Fig. 8.2. An incident neutron, travelling along the y-direction, interacts with several identical regions in which there is a static magnetic field B, applied along the x-direction. We describe the interaction by the Hamiltonian $H = \mu B \sigma_1$ [μ being the (modulus of the) neutron magnetic moment, and σ_1 the first Pauli matrix]. Let the initial neutron state be $\rho_0 = \rho_{\uparrow\uparrow} \equiv |\uparrow\rangle\langle\uparrow|$, where $|\uparrow\rangle$ and $|\downarrow\rangle$ are the spin states of the neutron along the z axis, which can be identified with the undecayed and decayed states of Section *2.3*, respectively. Assume that there are N regions in which B is present and that the interaction between the neutron and the magnetic fields has a total duration $T = N\tau$ ($\tau = \ell/v$, where ℓ is the

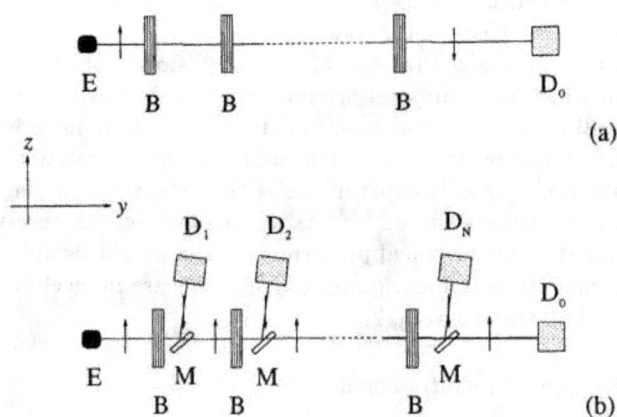

Fig. 8.2. (a) "Free" evolution of the neutron spin under the action of a magnetic field. An emitter E sends a spin-up neutron through several regions where a magnetic field B is present. The detector D_0 detects a spin-down neutron. (b) Quantum Zeno effect: the neutron spin is "monitored" at every step, by selecting and detecting the spin-down component. D_0 detects a spin-up neutron.

length of each region where B is present and v the neutron speed). It is then straightforward[153] to show that if T is chosen so as to satisfy the "matching" condition $T = (2m + 1)\pi/\omega$, where m is an integer and $\omega = 2\mu B/\hbar$, the final state and the probability that the neutron spin is found to be "down" at time T read respectively

$$\rho(T) = \rho_{\downarrow\downarrow}, \tag{8.39}$$

$$P_{\downarrow}(T) = 1. \tag{8.40}$$

The experimental setup in Fig. 8.2(a) is such that if the system is initially prepared in the up state, it will evolve to the down state after time T (π-pulse).

The situation outlined in Fig. 8.2(b) is very different. The experiment has been modified by inserting at every step a device able to select and detect the down component of the neutron spin. This is accomplished by a magnetic mirror M and a detector D. The magnetic mirror yields a spectral decomposition, by splitting a neutron wave with indefinite spin (a superposed state of up and down spins) into two branch waves (each of which is in a definite spin state along the z axis) and then forwarding the down component to the detector D. The action of the magnetic mirror can be compared to the inhomogeneous magnetic field in a typical Stern-Gerlach experiment. It is very important, in connection with the QZE, to bear in mind that the magnetic mirror does *not* destroy the coherence between the two branch waves: Indeed, the two branch waves corresponding to different spin states can be split coherently and brought back to interfere.[184,11] The global action of M and D is formally represented by the operator $\mathcal{O} \equiv \rho_{\uparrow\uparrow}$. If the initial neutron state and the "matching" condition for $T = N\tau$ are the same as in (8.39)-(8.40), the density matrix and the probability that the neutron spin is up at time $T = (2m + 1)\pi/\omega$ read respectively (in the notation of Section 2.3)

$$\rho^{(N)}(T) = V_N(T)\rho_0 V_N^{\dagger}(T) = \left(\cos^2 \frac{\omega\tau}{2}\right)^N \rho_{\uparrow\uparrow} = \left(\cos^2 \frac{\pi}{2N}\right)^N \rho_{\uparrow\uparrow}, \tag{8.41}$$

$$P_{\uparrow}^{(N)}(T) = \left(\cos^2 \frac{\pi}{2N}\right)^N. \tag{8.42}$$

This discloses the occurrence of a QZE: $P_{\uparrow}^{(N)}(T) > P_{\uparrow}^{(N-1)}(T)$ for $N \geq 2$, so that the evolution is "slowed down" as N increases. In the limit of infinitely many observations

$$\rho^{(N)}(T) \overset{N\to\infty}{\longrightarrow} \tilde{\rho}(T) = \rho_{\uparrow\uparrow}, \tag{8.43}$$

$$P_{\uparrow}^{(N)}(T) \overset{N\to\infty}{\longrightarrow} \mathcal{P}_{\uparrow}(T) = 1. \tag{8.44}$$

Frequent observations "freeze" the neutron spin in its initial state, by inhibiting ($N \geq 2$) and eventually hindering ($N \to \infty$) transitions to other states. Compare Eqs. (8.43) and (8.44) with (8.39) and (8.40): The situation is completely reversed.

It must be observed, however, that it is possible to obtain the *same* result without making use of projection operators, by simply performing a different analysis involving only unitary processes. Observe first that the preceding analysis involves only the neutron states. If the state of the total (neutron + detectors) system is taken into account, the initial state reads

$$\Xi_{\mathrm{I}}^{\mathrm{tot}} = (\xi_0 \otimes \rho_{\uparrow\uparrow}) \otimes \sigma_{\mathrm{I}}^{\mathrm{D_0}} \otimes \prod_{j=1}^{N} \sigma_{\mathrm{I}}^{\mathrm{D}_j}, \tag{8.45}$$

where $\xi_0 = |\phi_0\rangle\langle\phi_0|$, ϕ_0 being the neutron wave packet travelling along the horizontal direction in Fig. 8.2(b), and $\sigma_{\mathrm{I}}^{\mathrm{D}_j}$ the initial density matrix of detector D_j (displaying no result). By applying the quantum theory of measurement, it is straightforward to realize that the final state is

$$\Xi_{\mathrm{F}}^{\mathrm{tot}} = \left(\cos^{2N} \frac{\pi}{2N}\right) (\xi_0 \otimes \rho_{\uparrow\uparrow}) \otimes \sigma_{\mathrm{F}}^{\mathrm{D_0}} \otimes \prod_{j=1}^{N} \sigma_{\mathrm{I}}^{\mathrm{D}_j}$$

$$+ \sin^2 \frac{\pi}{2N} \sum_{k=1}^{N} \left(\cos^{2(k-1)} \frac{\pi}{2N}\right) (\xi_k \otimes \rho_{\downarrow\downarrow}) \otimes \sigma_{\mathrm{I}}^{\mathrm{D_0}} \otimes \sigma_{\mathrm{F}}^{\mathrm{D}_k} \otimes \prod_{j \neq k} \sigma_{\mathrm{I}}^{\mathrm{D}_j}, \tag{8.46}$$

where $\xi_k = |\phi_k\rangle\langle\phi_k|$, ϕ_k being the neutron wave packet travelling towards the kth detector, and $\sigma_{\mathrm{F}}^{\mathrm{D}_k}$ is the final density matrix of detector D_k (displaying neutron detection). Observe that in the above expression the total density matrix $\Xi_{\mathrm{F}}^{\mathrm{tot}}$ has no off-diagonal components as a consequence of the perfect measurement. The final total density matrix, in the channel representation, reads

$$\Xi_{\mathrm{F},ij}^{\mathrm{tot}} \equiv \begin{pmatrix} c^{2N}\rho_{\uparrow\uparrow} & & & & \text{\Large 0} \\ & s^2 c^{2N-2}\rho_{\downarrow\downarrow} & & & \\ & & s^2 c^{2N-4}\rho_{\downarrow\downarrow} & & \\ & & & \ddots & \\ \text{\Large 0} & & & & s^2\rho_{\downarrow\downarrow} \end{pmatrix},$$

$$(i, j = 0, 1, \ldots, N), \tag{8.47}$$

where $c \equiv \cos(\pi/2N)$ and $s \equiv \sin(\pi/2N)$. This corresponds to the case of frequent observations, in which the neutron route was observed at every step. The $i = j = 0$ component corresponds to detection by D_0, while the $i = j = n$ ($n = 1, \ldots, N$) component corresponds to detection in channel $N - n + 1$.

Remove now $D_1, \cdots, D_N$ in Fig. 8.2(b): In other words, do not perform any observation of the neutron route (except for the final measurement by D_0). We start from the initial state

$$\psi_{\mathrm{I}}^{\mathrm{tot}} = |\phi_0\rangle \otimes |\uparrow\rangle, \tag{8.48}$$

and a calculation analogous to the one sketched above yields the following final state:

$$\psi_{\mathrm{F}}^{\mathrm{tot}} = c^N |\phi_0\rangle \otimes |\uparrow\rangle + \left(-isc^{N-1}|\phi_N\rangle - isc^{N-2}|\phi_{N-1}\rangle - \ldots - is|\phi_1\rangle\right) \otimes |\downarrow\rangle. \tag{8.49}$$

The corresponding density matrix is (suppressing the spin state for simplicity)

$$\rho_{ij} \equiv \begin{pmatrix} c^{2N} & isc^{2N-1} & isc^{2N-2} & \cdots & isc^N \\ -isc^{2N-1} & s^2c^{2N-2} & s^2c^{2N-3} & \cdots & s^2c^{N-1} \\ -isc^{2N-2} & s^2c^{2N-3} & s^2c^{2N-4} & \cdots & s^2c^{N-2} \\ \vdots & \vdots & \vdots & \ddots & \vdots \\ -isc^N & s^2c^{N-1} & s^2c^{N-2} & \cdots & s^2 \end{pmatrix}, \quad (i,j = 0,1,\ldots,N). \tag{8.50}$$

The two expressions (8.47) and (8.50) clearly show that we have the *same* probability $P_\uparrow^{(N)} = [\cos^2(\pi/2N)]^N$ of detecting a spin-up neutron at D_0 *irrespectively* of the presence of detectors $D_1, \ldots, D_N$ in Fig. 8.2(b). It appears therefore that *no projection rule is necessary* in this context. The quantum Zeno effect can be given a purely dynamical explanation.

If $N \to \infty$, both density matrices tend to the limiting expression

$$\Xi_{\mathrm{F}}^{\mathrm{tot}}, \rho \longrightarrow \begin{pmatrix} \rho_{\uparrow\uparrow} & 0 & 0 & \cdots \\ 0 & 0 & 0 & \cdots \\ 0 & 0 & 0 & \cdots \\ \vdots & \vdots & \vdots & \ddots \end{pmatrix}. \tag{8.51}$$

This is the quantum Zeno paradox. It can be obtained by means of a purely dynamical process. The above conclusions, obtained for the particular experimental setup considered in this subsection, can be generalized to an arbitrary quantum system undergoing Zeno-type dynamics.[152] The hypotheses necessary in order to take the $N \to \infty$ limit ("continuous observation") will be discussed in Section *3.3*.

It must be stressed that an idea analogous to the one described in this subsection was outlined by Peres,[157] who made use of photons, rather than neutrons. His proposal inspired an interesting experiment.[110]

3.3   The $N \to \infty$ limit and its physical unrealizability

Is the so-called limit of continuous observation, as defined in (8.18)-(8.19), meaningful, from a physical point of view? Is it really possible to perform an infinite number of measurements in a finite elapse of time T? The beautiful theorem discussed in Section 2.3 is mathematically correct, but is also unphysical. It is easy to give physical counterexamples that show that the $N \to \infty$ limit cannot be taken.[131]

Let us discuss the experiment outlined in the previous subsection. First of all, the condition $\omega T = \omega N t = (2m + 1)\pi$, which is to be met at every step in Fig. 8.2(b), implies (by setting $m = 1$ for simplicity and without loss of generality)

$$B\ell = \frac{\pi \hbar v}{2\mu N} = O(N^{-1}), \tag{8.52}$$

where all quantities were defined before Eq. (8.39). It goes without saying that, as N increases, the practical realization of condition (8.52) becomes increasingly difficult. But there is even more to this: close scrutiny of Eq. (8.42) shows that $P_\uparrow^{(N)}(T)$ cannot tend to 1, even *in principle*, in the $N \to \infty$ limit, because of the uncertainty relations. Indeed, let ϕ be the argument of the cosine in Eq. (8.42)

$$\phi \equiv \frac{\pi}{2N} = \frac{\mu B\ell}{\hbar v}. \tag{8.53}$$

Mathematically, the above quantity is of order N^{-1}. On the other hand, from a *physical* point of view, it is impossible to avoid uncertainties in the neutron position Δx and speed Δv. As a consequence, ϕ cannot vanish, because it is lower bounded as follows

$$\phi = \frac{\mu B\ell}{\hbar v} > \frac{\mu B \Delta x}{\hbar v} > \frac{\mu B}{2Mv\Delta v}, \tag{8.54}$$

where M is the neutron mass and we assumed that the size ℓ of the interaction region (where the neutron spin undergoes a rotation under the action of the magnetic field) is larger than the longitudinal spread Δx of the neutron wave packet. The same bound holds in the opposite situation ($\ell < \Delta x$) as well.

By defining the magnetic energy gap $\Delta E_m = 2\mu B$ and the kinetic energy spread of the neutron beam $\Delta E_k = \Delta(Mv^2/2)$, the above inequality reads

$$\phi > \frac{1}{4} \frac{\Delta E_m}{\Delta E_k}. \tag{8.55}$$

It is now straightforward to obtain an expression for the value of the probability that a spin-up neutron is observed at D_0 when N is large:

$$P_\uparrow^{(N)}(T) \simeq (\cos\phi)^{2N} < \cos^{2N}\left(\frac{1}{4} \frac{\Delta E_m}{\Delta E_k}\right). \tag{8.56}$$

Notice that not only the above quantity does *not* tend to 1, but it *vanishes* in the $N \to \infty$ limit. In other words, in the experiment outlined in Fig. 8.2(b), *no* spin-up neutron would be observed at D_0 in the $N \to \infty$ limit!

What is reasonable to expect, in practice? The analysis of the previous subsections does not take into account the limits imposed by the uncertainty principle: As a matter of fact, N cannot be made arbitrarily large. To evaluate how large N can be in order that the QZE be observable in the above experiment, set $P_\uparrow^{(N)}(T) \sim 1/2$. We get

$$N \simeq \frac{16 \ln 2}{(\Delta E_m / \Delta E_k)^2} \sim 10^3, \tag{8.57}$$

for a thermal neutron ($\Delta E_k \simeq 75$ neV) in a typical magnetic field ($\Delta E_m \simeq 5$ neV) In conclusion, N turns out to be large enough in order that the QZE be experimentally observable, at least up to a certain approximation. We stress that in the above analysis we have neglected any losses and reflections[131] at the mirrors.

Criticisms against the physical meaning and realizability of the $N \to \infty$ limit were put forward some years ago by Ghirardi *et al.*[66] Although different from ours, these criticisms were based on the time-energy uncertainty relations. Other analyses of this problem can be found in Refs. 193 and 155.

4 Dynamical explanation of the experiment by Itano et al.

It is interesting to look at the experiment by Itano *et al.*,[84] outlined in Section *3.1*, in terms of the ideas of Section *3.2*. Notice that a purely dynamical explanation of this experiment was first proposed by Petrosky, Tasaki and Prigogine.[160] The initial state of the total system is

$$|\Psi_I\rangle = |\phi_1\rangle \otimes |0\rangle, \tag{8.58}$$

where $|\phi_i\rangle$ represents the atomic level #i ($i = 1, 2$) and $|0\rangle$ is the ground state of the Fock space (no photons).

The "free" evolution up to time τ simply yields Rabi oscillations between the atomic levels #1 and #2, as described in Section *3.1*,

$$|\Psi_I\rangle \to |\Psi_{(1)}\rangle = \left(\cos\frac{\omega\tau}{2}|\phi_1\rangle + i\sin\frac{\omega\tau}{2}|\phi_2\rangle\right) \otimes |0\rangle, \tag{8.59}$$

where ω is the frequency of the Rabi oscillations between levels #1 and #2. The $\omega_0 = (E_3 - E_1)/\hbar$ pulse yields in a very short time the following evolution

$$|\Psi_{(1)}\rangle \to |\Psi'_{(1)}\rangle = \cos\frac{\omega\tau}{2}|\phi_1\rangle \otimes |1\rangle + i\sin\frac{\omega\tau}{2}|\phi_2\rangle \otimes |0\rangle, \tag{8.60}$$

where $|1\rangle$ denotes a one-photon state. Notice that Eq. (8.60) is nothing but a spectral decomposition.[152] If N observations are performed during the time interval $(0, T)$, with $T = N\tau$, and if repopulation effects of the type described

in Section *3.1* are prevented, the survival probability for the atom in the first
level reads

$$\mathcal{P}_1^{(N)}(T) = \cos^{2N}\frac{\omega\tau}{2} = \cos^{2N}\frac{\pi}{2N}, \tag{8.61}$$

in agreement with (8.38).

5 A few comments

The main purpose of the present chapter is to discuss the quantum Zeno effect
and suggest a purely dynamical explanation, that does not involve any projection operators. The rather naive belief that a quantum-Zeno type dynamics
provides experimental evidence for the collapse of the wave function and supports the naive Copenhagen interpretation appears therefore completely un-
justified: The Schrödinger equation alone can yield a satisfactory explanation
of the QZE.

We repeat, once again, that a "projection" does not correspond to any
physical operation, and therefore should be regarded only as a convenient ex-
pedient (a "working rule") in order to account for the loss of quantum mechanical coherence (the collapse of the wave function). In this sense, the projection
rule is to be considered as a purely mathematical tool and no physical meaning
should be ascribed to it.

It is very difficult, in our opinion, to maintain the philosophical position
that a quantum measurement is a very peculiar physical process, that involves
a "classical" apparatus (whatever that means), many universes or sentient beings. We are forced to view the measurement process as a dynamical process
of some sort. It is nothing but a physical interaction, that takes place during
a certain elapse of time and follows (as far as we know) the laws of quantum
mechanics. Von Neumann's projection rules should never be applied lightheartedly, because by virtue (and in spite) of their effectiveness, they may lead
physicists astray.

One may wonder why one should endeavour to find a dynamical explanation for the QZE, rather than being content with a description in terms of
projection operators. The answer, which is also the leading philosophy of this
book, is that the projection rule is *no* explanation. It is an artificial technique
that is extraneous to quantum mechanics as a *physical* theory. An explanation
involving only dynamical laws is more satisfactory and does not require any
additional postulates. In this sense, it follows the prescriptions of Occam's ra-
zor. Moreover, a dynamical explanation contemplates the possibility that the
quantum coherence can be recovered, which would be just impossible within
the framework of the projection rule formalism.

It is also worth emphasizing that it can be difficult, in general, to obtain
an estimate for the characteristic time τ_Z introduced in (8.6). For example,

several authors considered the possibility that the proton decay has never been observed because its Zeno time τ_Z^{prot} might be longer than the lifetime of the Universe.[101,35,60,82,19]

Finally, we emphasize that the original ideas on the QZE involved truly *unstable* systems[21,100,59,46] rather than systems undergoing oscillations of some sort. Concerning this, as we have seen in Section *3.1*, one should prevent re-population effects on oscillating systems. A recent proposal[124] studies the temporal behaviour of a three-level atomic system that (unlike in the experiment by Itano *et al.*) is initially prepared in an excited state and undergoes a "dominated" evolution,[65,176] i.e. a "measurement" of some sort. In short, the situation considered is the following: Prepare a three-level atom in the excited level #2. The atom, if left undisturbed, would naturally decay to its ground state (assumed to be level #1) according to an (approximately) exponential law. Suppose now that you shine on the atom a very intense laser beam whose frequency is approximately equal to $(E_3 - E_1)/\hbar$. The atom, being continuously "monitored" by the laser beam, undergoes a sort of "continuous observation," in the following sense: If the electron in the atom makes a transition from level #2 to level #1, it is rapidly "snatched" away to level #3, so that any evidence that level #3 is populated would mean that level #2 has decayed. Level #2 is therefore continuously "observed" and we intuitively expect its decay to be hindered. This expectation is corroborated by a detailed calculation in Ref. 124. It concerns the possibility of observing a quantum-Zeno type effect even when the system has settled in the exponential region.

Chapter 9

QUANTUM DEPHASING BY CHAOS

In this chapter, we examine whether the chaotic behaviour of classical systems with a limited number of degrees of freedom can produce quantum dephasing. This is in contrast with the conventional idea that dephasing takes place only in large systems with a huge number of constituents and complicated internal interactions. On the basis of this analysis, we briefly discuss the possibility of defining quantum chaos and of inventing a "chaotic detector."

1 Quantum dephasing and chaos

As has been explained in this book, "quantum dephasing" is a central issue in quantum measurements. Many physicists used to think that quantum dephasing takes place as a result of the interaction with large systems, endowed with a huge number of elementary constituents and complicated internal interactions. Usually, some randomness or irreversibility associated with large systems is considered to play an important role in this context.

On the other hand, it is now known that a class of classical systems with *a few degrees of freedom* behaves chaotically due to nonlinear dynamics.[113] This leads us to expect that the interaction between a quantum particle and a classical system with a limited number of degrees of freedom, if the latter is in chaotic motion, may give rise to quantum dephasing on the former, against the above-mentioned naive expectation. In this chapter, we shall investigate the possibility of such a kind of mechanism (yielding quantum dephasing by the chaotic behaviour of a classical system with a few degrees of freedom), on the basis of simple model calculations. See Ref. 134.

Recently, Saito[173] suggested that we can talk about or define the (classical or quantum mechanical) chaotic behaviour of a dynamical system, through the observation of a possible randomization of the phase of a quantum particle interacting with the dynamical system. Our motive for the above investigation originated from his idea. As a natural extension of this argument, we shall also discuss the issue of quantum chaos,[a] and look for a way to invent a new type of "chaotic" detector. See Fig. 9.1. Note that the present discussion is somewhat tentative and needs further investigation. For simplicity, let us only discuss the

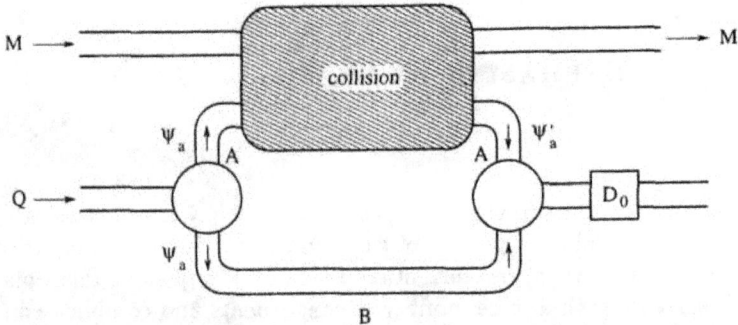

Fig. 9.1. Definition of the chaotic state of M by means of its dephasing effect on a quantum mechanical test particle Q.

case of a yes-no experiment, such as the Stern-Gerlach experiment. Refer to Fig. 1.3 or 1.4, and use the same notation as in Section *1.3* of Chapter 1. Recall that the whole measurement process is usually decomposed into two steps: the first being the spectral decomposition and the second the genuine detection. Each incoming particle emitted by E and represented by a wave packet ψ_0 is separated by V into two branch waves ψ_a and ψ_b, running in channels A and B, respectively, corresponding to mutually exclusive measurement propositions $\mathcal{A}$ and $\mathcal{B}$. In this spectral-decomposition step, the phase correlation between the two branch waves is fully kept. This is followed by the detection step at detector D_a placed in channel A. Suppose that ψ_a is changed to ψ'_a by passing through D_a. If we have a coincidence (or an anti-coincidence) signal between E and D_0, we get an affirmative (or negative) answer to $\mathcal{A}$. Here we have taken D_a to be a perfect detector, in the sense that it completely destroys the phase correlation between the two branch waves ψ_a and ψ_b, that is, it yields perfect dephasing between them.

Suppose that D_a is simply an instrument but not necessarily a perfect detector. In this case, in general, we are faced with a partially decohered case, as will presently be seen. If the two branch wave packets are guided into the final channel 0 to make a superposed state $\psi = \psi'_a + \psi_b$ (for simplicity, we have used the same notation for channel 0 as in channels A and B), the probability

---

[a]The readers can also refer to another paper, Ref. 10.

of observing the particle by a perfect detector D_0 placed in channel 0, reads

$$P = \int_\Omega |\psi|^2 dx = (\psi'_a, \psi'_a) + (\psi_b, \psi_b) + 2\text{Re}(\psi_b, \psi'_a), \tag{9.1}$$

where Ω is the non-vanishing support of the wave packets. Usually, we perform a quantum mechanical observation by sending many incoming particles (their total number, in an experimental run, being N_p) through a steady (and very weak) incident beam into the target (D_a in this case), and by accumulating many results obtained in such a run. If D_a is a perfect detector, that is, if it gives perfect dephasing between ψ'_a and ψ_b, the third term (the interference term) of (9.1) disappears for the accumulated distribution. If, on the other hand, D_a fully keeps the phase correlation between the branch waves, D_0 will read a perfect interference pattern. Otherwise, we encounter an imperfect measurement, yielding a partially decohered case. In this way one can see, by making use of D_0, whether D_a works well or not as a quantum detector.

In Chapter 4, we have introduced the *decoherence* parameter ϵ, defined by (4.11), to judge whether an instrument, D_a in this case, works well as a detector. We know that we have a perfect measurement, a perfect interference and an imperfect measurement, respectively, corresponding to the maximum ($= 1$), minimum ($= 0$) and intermediate values of the decoherence parameter. We are now going to estimate the value of the decoherence parameter for an instrument D_a containing a classical chaotic system.

For simplicity, let us assume that D_a consists of *only one* particle subject to classical dynamics, and that this classical particle (to be called D particle) interacts with an incoming quantum particle via a potential. The position of the D particle is the center of the potential; furthermore, all recoil effects and the internal structure of the D particle are neglected. In such a case, the motion of the ℓth incoming particle is described by the potential $\hat{V}_{(\ell)} \equiv V(\hat{r} - r_{(\ell)}(t))$, where $r_{(\ell)}$ is the position of the D particle when it meets the ℓth incoming particle. Our crucial assumption is that the D particle moves chaotically. Because of the ℓ-dependence of the potential center, we have to add the subscript (ℓ) to P and ψ'_a in the above expression (9.1). As the scatterer consists of a single particle and is not expected to be large in size, it is not appropriate to treat our present problem as an approximately one-dimensional collision process. This means that we cannot formulate the decoherence parameter in terms of the transmission coefficient T as has been done in Section *1.2* of Chapter 4. However, we will show below that, even in this case, it is still possible to formulate the decoherence parameter along a similar line of thought. The new decoherence parameter preserves its original interpretation: If dephasing takes place, the decoherence parameter is equal to unity and instrument D_a is considered to work well as a quantum detector.

In what follows, for simplicity, we suppress the subscript a of the branch wave. Of course, we shall keep the subscript (ℓ) for the branch wave of the ℓth incoming particle in channel A. We can write down the Schrödinger equation of the ℓth wave packet running in channel A as

$$\left[-\frac{\hbar^2}{2m}\nabla^2 + V(r - r_{(\ell)}(t))\right]\psi_{(\ell)}(r,t) = i\hbar\frac{\partial}{\partial t}\psi_{(\ell)}(r,t). \tag{9.2}$$

Note again that $r_{(\ell)}(t)$ stands for the center of the potential at time t. It is easy to see that the translation operator $\exp[(i/\hbar)\hat{p}\cdot r_{(\ell)}(t)]$ reduces (9.2) to

$$\left[-\frac{\hbar^2}{2m}\nabla^2 + V(r) + i\hbar v_{(\ell)}\cdot\nabla\right]\overline{\psi}_{(\ell)}(r,t) = i\hbar\frac{\partial}{\partial t}\overline{\psi}_{(\ell)}(r,t), \tag{9.3}$$

where $v_{(\ell)} = \dot{r}_{(\ell)}$ and

$$\overline{\psi}_{(\ell)}(r,t) = \exp[\frac{i}{\hbar}\hat{p}\cdot r_{(\ell)}(t)]\psi_{(\ell)}(r,t). \tag{9.4}$$

Here we shall consider only two extreme cases: (i) adiabatic change case, and (ii) rapid change case.

1.1 Adiabatic change

If the center of the potential moves very slowly during the passage of each incoming wave packet, we can safely consider the center fixed in each scattering process. In this case we can put $v_{(\ell)} = 0$, so that

$$\left[-\frac{\hbar^2}{2m}\nabla^2 + V(r)\right]\overline{\psi}_{(\ell)}(r,t) = i\hbar\frac{\partial}{\partial t}\overline{\psi}_{(\ell)}(r,t). \tag{9.5}$$

Here $\overline{\psi}_{(\ell)}$ becomes the wave function for the scattering process by a fixed potential $\hat{V}_0 \equiv V(r)$, whose center is located at the origin, so that we can omit the subscript (ℓ). (Recall that we are excluding the case in which the inner motion of the scatterer may give rise to an additional ℓ-dependence.) If we deal with a wave packet very close to a plane wave, we can put

$$\psi_{(\ell)}(r,t) \simeq \exp(-\frac{i}{\hbar}E_k t)u^{(+)}_{k(\ell)}(r), \tag{9.6}$$

$$\overline{\psi}_{(\ell)}(r,t) \simeq \exp(-\frac{i}{\hbar}E_k t)\overline{u}^{(+)}_{k}(r), \tag{9.7}$$

during the passage of the wave packet, where $E_k = \hbar^2 k^2/2m$ and

$$u^{(+)}_{k(\ell)}(r) = \exp(-\frac{i}{\hbar}\hat{p}\cdot r_{(\ell)})\overline{u}^{(+)}_{k}(r). \tag{9.8}$$

Here $u_{k(\ell)}^{(+)}(r)$ and $\bar{u}_k^{(+)}(r)$ are, respectively, the outgoing solutions of

$$\left[-\frac{\hbar^2}{2m}\nabla^2 + V(r - r_{(\ell)})\right] u_{k(\ell)}^{(+)}(r) = E_k u_{k(\ell)}^{(+)}, \tag{9.9}$$

$$\left[-\frac{\hbar^2}{2m}\nabla^2 + V(r)\right] \bar{u}_k^{(+)}(r) = E_k \bar{u}_k^{(+)}(r). \tag{9.10}$$

Note that $r_{(\ell)}$ is independent of time and that $\bar{u}_k^{(+)}$ has an $r_{(\ell)}$-dependent constant phase in order to match the boundary conditions for $u_{k(\ell)}^{(+)}$ and $\bar{u}_k^{(+)}$, both of which are subject to the plane wave normalization. Taking into account that $u_{k(\ell)}^{(+)} = \hat{W}_{(\ell)} u_k^{(0)}$ and $\hat{T}_{(\ell)} = \hat{V}_{(\ell)} \hat{W}_{(\ell)}$ ($\hat{W}_{(\ell)}$ and $\hat{T}_{(\ell)}$ being W- and T-matrices for the potential $\hat{V}_{(\ell)}$, respectively), we can write down the scattering amplitude as

$$\begin{aligned} F_{(\ell)}(k', k) &= -\frac{4\pi^2 m}{\hbar^2} (u_{k'}^{(0)}, \hat{T}_{(\ell)} u_k^{(0)}) \\ &= -\frac{4\pi^2 m}{\hbar^2} \exp[-iK \cdot r_{(\ell)}(t)](u_{k'}^{(0)}, \hat{T}_0 u_k^{(0)}), \end{aligned} \tag{9.11}$$

where $K = k' - k$ is the momentum transfer and $\hat{T}_0$ the T-matrix corresponding to the potential $\hat{V}_0$, because

$$\hat{V}_{(\ell)} = V(r - r_{(\ell)}(t)) = \exp[-\frac{i}{\hbar}\hat{p} \cdot r_{(\ell)}(t)]\hat{V}_0 \exp[\frac{i}{\hbar}\hat{p} \cdot r_{(\ell)}(t)], \tag{9.12}$$

$$\hat{T}_{(\ell)} = \exp[-\frac{i}{\hbar}\hat{p} \cdot r_{(\ell)}(t)]\hat{T}_0 \exp[\frac{i}{\hbar}\hat{p} \cdot r_{(\ell)}(t)]. \tag{9.13}$$

These formulae are easily understood, for example, on the basis of the Born approximation

$$\begin{aligned} F_{(\ell)}(k', k) &\simeq -\frac{m}{2\pi\hbar^2} \int d^3r\, e^{-iK \cdot r} V(r - r_{(\ell)}(t)) \\ &= -\frac{4\pi^2 m}{\hbar^2} e^{-iK \cdot r_{(\ell)}(t)} (u_{k'}^{(0)}, \hat{V}_0 u_k^{(0)}), \end{aligned} \tag{9.14}$$

and its generalization

$$\hat{V}_0 \longrightarrow \hat{T}_0 = \hat{V}_0 + \hat{V}_0 \frac{1}{E_k - \hat{H} + i\epsilon} \hat{V}_0, \tag{9.15}$$

with the total Hamiltonian $\hat{H} = \hat{p}^2/2m + \hat{V}_0$. If we are dealing with low energy scattering by a short-distance force, we can put

$$F_{(\ell)}(k', k) = -kb \exp[-iK \cdot r_{(\ell)}], \tag{9.16}$$

where b is the scattering length. Note that usually, we can neglect the ℓ-dependence of the scattering length.

Along the general line of thought given in Section 1.2 of Chapter 4, we introduce the *decoherence* parameter in a three-dimensional scattering process

$$\epsilon \equiv 1 - \frac{\left|\int_{\Delta\omega} d\omega \overline{F}\right|^2}{\int_{\Delta\omega} d\omega |F|^2} = 1 - \left|\frac{1}{\Delta\omega}\int_{\Delta\omega} d\omega \ \overline{\exp[-i\mathbf{K}\cdot\mathbf{r}_{(\ell)}]}\right|^2, \qquad (9.17)$$

where $\Delta\omega = (\Delta\theta, \Delta\varphi)$ stands for the solid angle around $\omega_0 = (\theta_0, \varphi_0)$ under which the scatterer sees the detector. Clearly this ϵ provides a quantitative measure of the degree of quantum dephasing, as in the one-dimensional case. Notice that this ϵ depends on θ_0 and φ_0, in general.

In order to estimate $\overline{\exp[-i\mathbf{K}\cdot\mathbf{r}_{(\ell)}]}$, consider that $\mathbf{r}_{(\ell)}$ is the resultant point of a random walk, $\mathbf{r}_{(1)} \to \mathbf{r}_{(2)} \to \cdots$, according to the recent study of classical chaos.[113] Therefore, if the incident beam is steady and very weak, we can use the Gaussian law with characteristic length ΔL for the distribution of $\mathbf{r}_{(\ell)}$, or, in other words, we can replace the above bar-averaged quantity $\overline{\exp[-i\mathbf{K}\cdot\mathbf{r}_{(\ell)}]}$ with

$$\exp[-N_p\frac{(\Delta L)^2}{2}K^2] = \exp[-N_p(\Delta L)^2 k^2(1-\cos\theta_0)], \qquad (9.18)$$

where $K = 2k\sin(\theta_0/2)$.

The angle integrals are computed in the following way:

$$\int_{\Delta\omega} d\omega = \Delta\varphi[\cos\theta_0 - \cos(\theta_0 + \Delta\theta)]$$

$$\simeq \Delta\varphi[\Delta\theta\sin\theta_0 + \frac{1}{2}(\Delta\theta)^2\cos\theta_0], \qquad (9.19)$$

$$\int_{\Delta\omega} e^{-N_p(\Delta L)^2 k^2(1-\cos\theta)}d\omega = \Delta\varphi e^{-N_p(\Delta L)^2 k^2}\int_{\cos(\theta_0+\Delta\theta)}^{\cos\theta_0} e^{N_p(\Delta L)^2 k^2\xi}d\xi$$

$$\simeq \frac{\Delta\varphi\, e^{-N_p(\Delta L)^2 k^2(1-\cos\theta_0)}}{N_p(\Delta L)^2 k^2}\left(1 - e^{-N_p(\Delta L)^2 k^2[\Delta\theta\sin\theta_0+\frac{1}{2}(\Delta\theta)^2\cos\theta_0]}\right).$$

$$(9.20)$$

Thus we obtain

$$\epsilon \simeq 1 - \left(\frac{e^{-N_p(\Delta L)^2 k^2(1-\cos\theta_0)}\left(1 - e^{-N_p(\Delta L)^2 k^2[\Delta\theta\sin\theta_0+\frac{1}{2}(\Delta\theta)^2\cos\theta_0]}\right)}{N_p(\Delta L)^2 k^2[\Delta\theta\sin\theta_0 + \frac{1}{2}(\Delta\theta)^2\cos\theta_0]}\right)^2,$$

$$(9.21)$$

from which we conclude that

$$\epsilon \simeq 0 \quad \text{for} \quad N_p(\Delta L)^2 k^2 [\Delta\theta \sin\theta_0 + \frac{1}{2}(\Delta\theta)^2 \cos\theta_0] \ll 1, \tag{9.22}$$

and

$$\epsilon \simeq 1 \quad \text{for} \quad N_p(\Delta L)^2 k^2 [\Delta\theta \sin\theta_0 + \frac{1}{2}(\Delta\theta)^2 \cos\theta_0] \gg 1. \tag{9.23}$$

We get coherence in the case (9.22), and dephasing in the case (9.23).

Consequently, we arrive at the conclusion that the chaotic motion of a classical system can generate dephasing for sufficiently large ΔL, even though the system has very few degrees of freedom, or alternatively, when the classical system has reached a well-developed ("aged") stage, i.e. $N_p \gg 1$ so that the replacement (9.18) becomes quite reasonable, after the interaction with many incoming particles. This suggests the possibility of inventing a new type of quantum detector, by making use, for instance, of a "randomly oscillating mirror." On the contrary, we observe coherence for very small ΔL, or when the system is in the developing stage, before "aging" ($N_p \simeq 1$), even though in the latter case we have no sound reasoning to justify the replacement (9.18).

1.2 Rapid change

Let us consider the case in which $r_{(\ell)}(t)$ in (9.2) changes very rapidly, during the passage of the ℓth incoming wave packet. In this case, we may first replace the potential term with

$$\overline{V(r - r_{(\ell)}(t))} \equiv \frac{1}{\tau} \int_t^{t+\tau} dt V(r - r_{(\ell)}(t))$$
$$= \int d^3r' \, V(r - r') w_{(\ell)}(r'), \tag{9.24}$$

$$w_{(\ell)}(r') \equiv \frac{1}{\tau} \int_t^{t+\tau} dt \delta(r' - r_{(\ell)}(t)), \tag{9.25}$$

where τ is the passage time of the wave packet. Furthermore, let us restrict ourselves to the situation in which $w_{(\ell)}(r')$ can be replaced with a statistical distribution $\mathcal{W}_{(\ell)} = \mathcal{W}(r' - R_{(\ell)})$, which has width $\sim \Delta R_{(\ell)}$ around $R_{(\ell)}$, and has no time dependence except that through ℓ. Under these circumstances, we can practically reduce our scattering problem to that of an average potential given by

$$\mathcal{V}(r - R_{(\ell)}) = \int V(r - r') \mathcal{W}(r' - R_{(\ell)}) d^3r', \tag{9.26}$$

leading to the effective Schrödinger equation

$$\left[-\frac{\hbar^2}{2m} \nabla^2 + \mathcal{V}(r - R_{(\ell)}) \right] \psi_{(\ell)}(r,t) = i\hbar \frac{\partial}{\partial t} \psi_{(\ell)}(r,t). \tag{9.27}$$

It is remarked that in using this equation we are neglecting some sort of higher-order fluctuation effects on the Schrödinger wave function.

For the particular case $V(r) = (2\pi\hbar^2/m)b\delta(r)$ (i.e. Yang's approximation, b being the scattering length), we further reduce (9.26) to

$$V(r - R_{(\ell)}) = \frac{2\pi\hbar^2}{m}bW(r - R_{(\ell)}), \qquad (9.28)$$

as a good approximation, because the force range of V is much shorter than $|\Delta R_{(\ell)}|$.

We have now found a parallelism between the adiabatic change case and the rapid change case in the above approximation, with correspondence between $r_{(\ell)}$ and $V(r - r_{(\ell)})$ [see (9.2)] in the former case, and $R_{(\ell)}$ and $V(r - R_{(\ell)})$ in the latter case, respectively. Note that $R_{(\ell)}$ and $r_{(\ell)}$ are assumed constant in both cases. Therefore, we can extend the arguments on the conditions for quantum dephasing in the adiabatic change case to the present case as well.

It should be remarked that $W_{(\ell)}$, in many practical cases, describes a very dilute and broad distribution, in which we can regard (9.28) as a constant potential with strength $(2\pi\hbar^2/m)bW_{(\ell)}(0)$ $(W_{(\ell)}(0) \simeq (|\Delta R_{(\ell)}|)^{-3})$ over a wide spatial region of (large) spread $|\Delta R_{(\ell)}|$. In this case, we can easily estimate the scattering phase shift (χ) by

$$\chi = -\frac{\lambda b}{|\Delta R_{(\ell)}|^2}. \qquad (9.29)$$

Here λ is the particle wavelength and $\rho \simeq |\Delta R_{(\ell)}|^{-3}$ the scatterer density. For very large $|\Delta R_{(\ell)}|$, this phase shift becomes very small. This means that we can hardly observe quantum dephasing in this case. In conclusion, this type of instrument is nothing but a phase shifter, which can never yield quantum dephasing.

1.3 Remarks

We have so far discarded possible effects of scatterer recoil. In order to take these recoil effects into account, we just have to reformulate the scattering amplitude in the above discussion, in an appropriate way, according to quantum theory of scattering. In this way, we can discuss the following two possibilities.

Quantum chaos: Consider the case in which both incoming and target particles are quantum mechanical. (Note that the target particle has been treated as a classical particle in the above discussion.) The formulation described above still holds if we use the quantum mechanical scattering amplitude (including the recoil effect) for the collision between incoming and target particles. Within this framework, we may be able to obtain the notion of "quantum chaos," for the *target particle state*, via the observation of "quantum dephasing" on the scattering amplitude of the incoming particle. See Fig. 9.1. As was already mentioned, Saito[173] suggested that quantum chaos can arise from possible random phases of the quantum mechanical scattering amplitude in the path-integral representation. This idea may be concretized by replacing the T-matrix in formula (9.11) with related quantities in the path-integral representation, in particular with those in the WKB approximation. This, as a natural extension of the present approach, is a promising means to open a doorway into quantum chaos.

Chaotic detector: If we can detect the above-mentioned recoil effect as a signal, we have the possibility of making detectors that contain only a few constituents, for example, by means of a randomly moving mirror. Detectors of this kind are quite new; conventional detectors have a huge number of constituents. Even though we know that the generation and detection of such a signal would pose a difficult problem, one of such possibilities would be to utilize the Fourier analysis of the response functions in momentum space. If the detector is characterized by a large value of the decoherence parameter ϵ, we can extract the signal information by observing the Fourier spectrum of the correlation of the wave functions, defined by $(\psi_b, \hat{I}(k)\psi'_a)(t)$, where $\hat{I}(k)$ stands for a spectral function selecting the momentum components around k. See the discussion at the end of Section *1.2* of Chapter 4. We expect such a correlation function to depend strongly on ϵ, in particular when its value is close to unity (dephasing). The realization of this kind of detector will become one of the most interesting and important issues in the future.

OUTLOOK

In this book we have discussed the quantum measurement problem, concentrating our attention on the many-Hilbert-space theory. This theory takes into account the macroscopic structure of a measuring apparatus, analyzing the whole measurement process in quantum mechanical terms. This is an important characteristic of this approach: No "classical" behaviour is postulated for macroscopic objects. Instead, they are always treated quantum mechanically. The leading philosophy is that the classical approximation should be *proven* to be a good one, rather than being simply postulated. This is one of the most important differences with the naive Copenhagen interpretation. As we have seen in Chapter 4, the many-Hilbert-space theory describes classical properties by means of a continuous superselection rule, which makes possible the transition to a "large" Hilbert space, in which one can define "classical" observables, belonging to the center of von Neumann's algebra.

The many-Hilbert-space theory enables one to discuss imperfect measurement via the introduction of the so-called decoherence parameter ϵ. This parameter, introduced and discussed in Chapter 4, has been utilized and evaluated in the subsequent chapters. It takes values in the range $[0, 1]$: If $\epsilon = 0$, the quantum mechanical coherence is fully preserved and one is able to observe typical quantum effects, such as interference. On the other hand, if $\epsilon = 1$, the quantum coherence is completely lost, and the quantum effects typically disappear; for example, a perfect quantum measurement is always characterized by the value $\epsilon = 1$. The intermediate values $0 < \epsilon < 1$ reflect physical situations in which the quantum coherence is only partially lost. A typical example is an imperfect measurement, where some residual quantum coherence can be brought to light by a suitable experimental technique.

The intermediate values of the decoherence parameter $0 < \epsilon < 1$ enable one to interpret mesoscopic phenomena in terms of imperfect measurements. In a typical mesoscopic system the quantum coherence is partially preserved: the evaluation of the decoherence parameter in this case is an interesting open problem. The value of ϵ might lead to quantitative estimates of the "degree of mesoscopicity" of a given system.

It must be noted that the loss of quantum coherence characterized by $\epsilon < 1$

is always irretrievable. This is to be contrasted with other typical situations in which the lack of spatial overlap of the wave packets results in a vanishing interference. This case has been discussed in Chapter 6, in relation to neutron interferometry. As we have seen there, a realistic evaluation of the decoherence parameter, in this situation, displays a large degree of quantum coherence. The lack of spatial overlap of the wave packets does not necessarily imply a loss of quantum mechanical coherence. This is an important idea, both on experimental and theoretical grounds.

A summary of the properties of ϵ and the phenomena it characterizes is given in Table 1.

Table 1: The decoherence parameter ϵ.

Value of ϵ	Coherence properties	Typical phenomena
$\epsilon = 0$	perfect coherence	perfect interference
$0 < \epsilon < 1$	partial loss of coherence (partial decoherence)	imperfect measurements partial dephasing mesoscopic systems nondemolition measurements
$\epsilon = 1$	complete loss of coherence (complete decoherence)	quantum measurements total dephasing "classical" behaviour

The many-Hilbert-space approach enables one to analyze quantitatively real experiments, in diverse areas of physics. This is to be contrasted to many other existing theories, that are rather confined to a philosophical or a very theoretical domain. We have seen several examples of application to real experiments, in this book. In some cases, it is possible to explain experimental data that so far seemed difficult to understand. One typical example has been discussed in Chapter 6, Section 2. The application of the many-Hilbert-space theory and the evaluation of the decoherence parameter in some theoretical models and in numerical simulations was shown in Chapters 5 and 7, respectively. As we have seen, there are still open problems, awaiting possible applications.

It is important to realize that the many-Hilbert-space approach endeavours to describe the measurement process and the decoherence phenomenon within quantum mechanics, by making use of the Schrödinger equation. The basic quantum equations and the fundamental quantum mechanical framework are never modified. Moreover, no additional "hidden" variable is introduced. We always follow the quantum mechanical prescriptions. Furthermore, we have endeavoured to describe and define new phenomena (such as quantum chaos—see Chapter 9) by starting from quantum mechanical properties, such as the phase of the wave function. This has led us to consider new exciting possibilities, such as that of conceiving a "chaotic detector."

There are many open problems, to be investigated in the future. Quantum chaos, mesoscopic physics, new neutron interferometric experiments and the so-called quantum non-demolition measurements are some of these. In some sense, one of the virtues of the many-Hilbert-space approach is just its ability to analyze quantitatively new phenomena and enable one to make research in new areas of physics. In this context, we stress that this approach has, so to speak, "pulled" the quantum measurement problem back to real physics, after many years of academic and rather philosophical debate. Even though we are by no means against philosophical discussions, we also think that the quantum measurement problem, together with some subtle conceptual issues related to decoherence and mesoscopic physics, have led the foundations of quantum mechanics to the very frontiers of physics. The best proof in support of this claim can be seen in the growing number of international conferences on related topics and the discovery of many applications in high technology.

However, we should not forget that quantum mechanics itself is a young theory. As a matter of fact, it was born in 1925, which is the year of birth of one of the present authors! In this sense, one should not expect that quantum mechanics and therefore the quantum measurement problem are firmly and "eternally" established. They both need much additional theoretical and experimental research.

In our opinion, quantum mechanics will soon be at a turning point. It is difficult, if not impossible, to foresee when and how this can happen, but we feel that this will not entail any "return" to classical physics. However, any drastic change in the quantum formalism will necessarily introduce a similarly drastic change in our comprehension of quantum measurements.

APPENDIX

A   Temporal behaviour of quantum systems

The temporal evolution of a quantum mechanical system, initially prepared in an eigenstate of the unperturbed Hamiltonian, is known to be roughly characterized by three distinct regions: A quadratic behaviour at short times, an approximate exponential decay at intermediate times, and a power law at long times. The temporal behaviour of quantum systems at very short times leads to the quantum Zeno effect, discussed in detail in Chapter 8. This appendix is devoted to the study of the temporal behaviour of quantum systems at longer times; this will enable us to focus on the status of the familiar exponential law in quantum theory.

An exhaustive analysis of the temporal behaviour of quantum mechanical systems, especially at very long times, is a rather complex problem. We shall only present a brief and elementary exposition of this issue; see Ref. 132 for a complete discussion and further references.

We assume that the total Hamiltonian can be written as $H = H_0 + H'$, where H_0 and H' are the unperturbed and the interaction Hamiltonians, respectively.[a] Let $|n\rangle$ be an eigenstate of H_0, with eigenvalue E_n

$$H_0|n\rangle = E_n|n\rangle, \qquad 1 = \sum_n |n\rangle\langle n|. \qquad (A.1)$$

Though we are using the discrete notation for the sake of simplicity, it should be understood that the extension to continuous cases will be done whenever necessary. The initial state of the system is taken to be $|a\rangle$, an eigenstate of H_0, and we switch on the interaction at $t = 0$. We also assume that the interaction Hamiltonian H' satisfies the conditions

$$\langle n|H'|n'\rangle = 0, \qquad \langle a|H'|n\rangle \neq 0, \qquad (n, n' \neq a), \qquad (A.2)$$

so that the only nonvanishing matrix elements are those between the initial state $|a\rangle$ and the other states $|n\rangle$ $(n \neq a)$.

---

[a]The caret on operators will be suppressed in this appendix.

The above assumptions are rather restrictive: One can require much weaker conditions. Notice, however, that the present discussion has only illustrative purposes, for which these assumptions are of great help. A few comments are in order: First, the first relation in (A.2) corresponds to mass renormalization in field theory and enables us to consider and formally solve the Lippmann-Schwinger equation

$$|\psi_a\rangle = |a\rangle + \frac{1}{E_a - H_0} H'|\psi_a\rangle. \tag{A.3}$$

Second, the same relation is a somewhat strong version of the so-called random phase approximation.[154] Finally, notwithstanding the conditions (A.2), the above can still be considered as a rather general interaction Hamiltonian, widely used in elementary textbooks (e.g., Ref. 123).

A.1 Short times

We shall work in the interaction picture and study the temporal behaviour of the survival probability amplitude $\langle a|U_I(t)|a\rangle$. The evolution operator in the interaction picture $U_I(t)$ satisfies the Schrödinger equation ($\hbar = 1$)

$$i\frac{d}{dt}U_I(t) = H'_I(t)U_I(t), \qquad H'_I(t) = e^{iH_0 t}H'e^{-iH_0 t}, \tag{A.4}$$

where $H'_I(t)$ is the interaction Hamiltonian in the interaction picture. The formal solution for $U_I(t)$ is recursively obtained from the integral equation

$$U_I(t) = 1 - i\int_0^t dt' H_I(t')U_I(t') \tag{A.5}$$

with the initial condition $U_I(0) = 1$. By making use of (A.2) one easily shows[123,132] that

$$\frac{d}{dt}\langle a|U_I(t)|a\rangle = -\int_0^t dt_1 e^{iE_a(t-t_1)}f(t-t_1)\langle a|U_I(t_1)|a\rangle, \tag{A.6}$$

where

$$f(t) = \langle a|H'e^{-iH_0 t}H'|a\rangle. \tag{A.7}$$

For small t, we just expand the r.h.s. of (A.6) in powers of t

$$\frac{d}{dt}\langle a|U_I(t)|a\rangle = -f(0)\langle a|U_I(0)|a\rangle t + O(t^2) \tag{A.8}$$

and obtain

$$\langle a|U_I(t)|a\rangle \simeq 1 - \frac{f(0)}{2}t^2. \tag{A.9}$$

Notice that $f(0) = \langle a|H'^2|a\rangle$ is positive definite. This is in complete agreement with the analysis of Chapter 8. Consider Eqs. (8.5) and (8.6), with $|a\rangle$ eigenstate of the free Hamiltonian H_0 and the interaction satisfying conditions (A.2). As we saw, this behaviour engenders the QZE.

A.2 Very long times

One might expect that even though it starts to evolve quadratically in time, as we have just seen, the survival probability amplitude $\langle a|U_{\rm I}(t)|a\rangle$ would show the familiar exponential decay at longer times $t \to \infty$, just as in classical mechanics. This naive guess turns out to be incorrect, as a consequence of some mathematical properties of the survival amplitude. Indeed, in terms of the total Hamiltonian H, the evolution operator in the interaction picture reads

$$U_{\rm I}(t) = e^{iH_0t}e^{-iHt} \qquad (A.10)$$

and by introducing a complete orthonormal set of eigenstates of H, we easily obtain the following expression for the survival probability amplitude

$$\langle a|U_{\rm I}(t)|a\rangle = e^{iE_a t}\int \omega_a(E)e^{-iEt}dE, \qquad (A.11)$$

where $\omega_a(E)$ is the energy density of the initial state

$$\omega_a(E) \equiv \sum_{E_n=E} |\langle n|a\rangle|^2. \qquad (A.12)$$

Here the summation is taken over all the quantum numbers, except energy, that are necessary for the specification of a complete orthonormal set. The survival probability amplitude is therefore the Fourier transform of the energy density $\omega_a(E)$, as was first pointed out by Fock and Krylov.[57] Let us now assume, on physical grounds, that the spectrum of the total Hamiltonian is bounded from below so that the vacuum state is stable: There exists a certain finite energy E_g below which the function $\omega_a(E)$ vanishes. Khalfin showed[99] that this condition on ω_a requires that its Fourier transform (i.e. the survival probability amplitude) satisfies the inequality

$$\int_{-\infty}^{\infty} \frac{|\ln|\langle a|U_{\rm I}(t)|a\rangle||}{1+t^2}dt < \infty, \qquad (A.13)$$

as a consequence of the Paley-Wiener theorem.[151] The inequality (A.13) implies that the survival probability cannot decay exponentially: The decay process is slower than any exponential at large times.

A.3 Exponential law at intermediate times: The Fermi Golden Rule

The above result might seem somewhat puzzling, but is an unavoidable consequence of mathematical properties of the evolution law: Strictly speaking, an exponential decay can neither be expected at very short, nor at very long times in quantum systems. On the other hand, however, we know that the

exponential law for decay processes has been well confirmed experimentally
and no deviation from it at long times has ever been observed. See Ref. 36
for a survey of experiments. How and where can one find a resolution to this
apparent contradiction?

In order to understand the status of the exponential law in quantum mechanics, we shall go back to the Schrödinger equation for the survival amplitude
(A.6). We follow the usual Wigner-Weisskopf Laplace transform method and
transform (A.6) into an algebraic equation (incorporating the initial condition),
to obtain

$$\langle a|U_I(t)|a\rangle = \frac{1}{2\pi i}\int_{-i\infty+\varepsilon}^{i\infty+\varepsilon}\frac{e^s}{g(s,t)}ds, \tag{A.14}$$

where

$$g(s,t) = s + t\int_0^\infty e^{-\frac{s}{t}u}e^{iE_a u}f(u)du. \tag{A.15}$$

This is the solution in a closed form.

Since the amplitude $\langle a|U_I(t)|a\rangle$ is given by the inverse Laplace transform
(A.14), we understand that the exponential decay occurs only when the complex function $g(s,t)$ has zeros in the left-half complex s plane. In order to
perform the inverse Laplace transform, the function g has to be analytically
continued into the left-half complex s plane. By performing the integration
over u in (A.15) we obtain

$$g(s,t) = s + t\langle a|H'\frac{i}{E_a - H_0 + i\frac{s}{t}}H'|a\rangle$$

$$= s + it\int_{C_0}\frac{\sum_r |\langle E_0, r|H'|a\rangle|^2}{E_a - E_0 + i\frac{s}{t}}dE_0. \tag{A.16}$$

Here $|E_0, r\rangle$ are eigenstates of H_0, and form a complete orthonormal set, r
being quantum numbers describing possible degeneracies. Observe that the
integrand has a simple pole at $E_0 = E_a + i\frac{s}{t}$. Since $t > 0$ and s is taken to
have a positive real part (in order to assure the convergence of the integration
over t in the Laplace transform), the pole lies above the integration contour C_0
which extends along the real E_0 axis. Therefore in order to extend the function
$g(s,t)$ into the left-half complex s plane, where the real part of s is negative,
the integration contour should be deformed in the complex E_0 plane so that its
relative configuration w.r.t. the singularity is maintained.[7] See Fig. A.1. This
gives us the following change for the last term in (A.16)

$$it\int_{C_0}\frac{\sum_r |\langle E_0, r|H'|a\rangle|^2}{E_a - E_0 + i\frac{s}{t}}dE_0$$

$$\longrightarrow it\int_{C_0}\frac{\sum_r |\langle E_0, r|H'|a\rangle|^2}{E_a - E_0 + i\frac{s}{t}}dE_0 + 2\pi t\sum_r |\langle E_0, r|H'|a\rangle|^2\Big|_{E_0 = E_a + i\frac{s}{t}}, \tag{A.17}$$

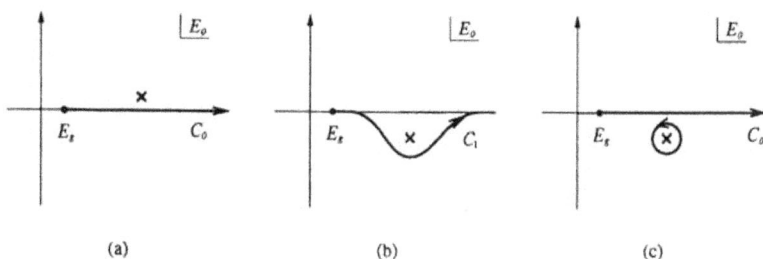

Fig. A.1. Integration contours (a) C_0 for $\text{Re}(s) > 0$ and (b) C_1 for $\text{Re}(s) < 0$. The former extends along the real E_0 axis starting from the lowest energy E_g. The pole at $E_0 = E_a + i\frac{s}{t}$ is located above the contours C_0 and C_1. (c) The contour C_1 can further be deformed and decomposed into the contour C_0 along the real E_0 axis and a circle surrounding the pole.

as the pole moves into the lower-half E_0 plane crossing the real E_0 axis. Notice that the second term of the r.h.s., which represents the contribution from the simple pole now appearing in the lower-half E_0 plane, has in general a nonvanishing real part. We see that this term just expresses the discontinuity of the l.h.s. over the imaginary s axis, thus ensuring the analyticity of the r.h.s. as a whole.

It is now clear that the analytically continued function $1/g(s,t)$ has the following properties as a function of the complex variable s:

- Under the assumption that H_0 (like H) has a bounded spectrum, a branch cut exists along the imaginary s axis, extending to $-i\infty$ from the branch point at $s = it(E_a - E_g)$, E_g being the lowest value of the spectrum of H_0.

- On the first Riemannian sheet, into which the analytic continuation must be done without crossing the branch cut, the last term in (A.17) does not show up. The function $1/g(s,t)$ has no singularity and is analytic in this plane.

- The last term in (A.17) appears only when the function $g(s,t)$ is continued into the second Riemannian sheet through the cut. This term allows the function $1/g(s,t)$ to have a simple pole in this plane where $\text{Re}(s) < 0$.

These properties of the function $1/g(s,t)$ are enough to understand that the survival probability amplitude $\langle a|U_I(t)|a\rangle$, being nothing but the inverse Laplace transform of $1/g(s,t)$, exhibits two distinct behaviours at large times. Observe that the integration contour over s in (A.14) can be deformed in the left-half plane as in Fig. A.2(b). Let $s_0 = -\frac{\gamma}{2}t - i\delta E t$ $(\gamma > 0)$ be the zero of

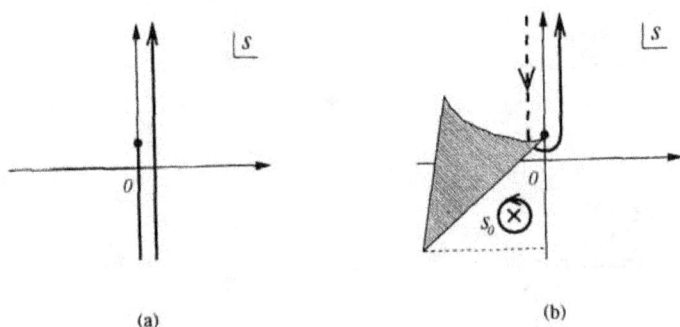

(a)   (b)

Fig. A.2. Integration contours over s. (a) The original one and (b) the deformed one. The latter is composed of a circle surrounding the simple pole at s_0 on the second Riemannian sheet and a path starting from $i\infty$ on the second sheet, turning around the branch point at $s = it(E_a - E_g)$ and extending back to $i\infty$ on the first sheet.

$g(s, t)$ in the second Riemannian sheet, so that

$$-\frac{\gamma}{2}\left[1 + \int_{E_g}^{\infty} \frac{\sum_r |\langle E_0, r|H'|a\rangle|^2}{|E_a - E_0 + \delta E - i\frac{\gamma}{2}|^2}dE_0\right]$$
$$+ 2\pi \sum_r |\langle E_0, r|H'|a\rangle|^2\Big|_{E_0 = E_a + \delta E - i\frac{\gamma}{2}} = 0 \quad (A.18)$$

and

$$-\delta E\left[1 - \int_{E_g}^{\infty} \frac{\sum_r |\langle E_0, r|H'|a\rangle|^2}{|E_a - E_0 + \delta E - i\frac{\gamma}{2}|^2}dE_0\right]$$
$$+ \int_{E_g}^{\infty} \frac{(E_a - E_0)\sum_r |\langle E_0, r|H'|a\rangle|^2}{|E_a - E_0 + \delta E - i\frac{\gamma}{2}|^2}dE_0 = 0. \quad (A.19)$$

Notice that both γ and δE, which are obtained as solutions of these equations, are t-independent constants. Close scrutiny[7] shows that in the weak-coupling limit, the second term in the square brackets of (A.18) gives a finite contribution and we exactly reproduce Fermi's Golden Rule,[56] according to which the decay rate is given by

$$\tau_E^{-1} \equiv \gamma = 2\pi \sum_r |\langle E_0, r|H'|a\rangle|^2 \quad (A.20)$$

by the first-order perturbation theory. We see again that the presence of the second term in the l.h.s. of (A.18) is crucial for $g(s, t)$ to have a zero in the left-half s plane. This simple-pole contribution to $1/g(s, t)$ yields the exponential decay form for the survival amplitude, while the integration along the imaginary s axis cancels the former to yield the quadratic behaviour at short times, according to the argument exposed in Section *A.1*. The latter contribution is also responsible for the deviation from the exponential law at very long times, as expected on the basis of the afore-mentioned theorem (A.13).

Summarizing, the exponential decay form can be expected only in intermediate times in quantum mechanics: The nonexponential behaviour at short and long times arises from the contribution of a dispersion integral along a branch cut in the complex energy plane. Such a branch cut starts from the lower bound of the energy spectrum. Exhaustive analyses on this subject can be found in Refs. 55, 58, 132.

Bibliography

[1] Alefeld, B., G. Badurek and H. Rauch, *Z. Phys.* **41B** (1981) 231.

[2] Alley, C.O., O.G. Jakubowicz and W.C. Wicks, in *Proc. Second International Symposium Foundations of Quantum Mechanics*, eds. M. Namiki *et al.* (Phys. Soc. Japan, Tokyo, 1987), p.36.

[3] Alley, C.O. and Y.H. Shih, *Phys. Rev. Lett.* **61** (1988) 2921.

[4] Altenmuller, T.P. and A. Schenzle, *Phys. Rev.* **A49** (1994) 2016.

[5] Alvarez, L.W. and F. Bloch, *Phys. Rev.* **57** (1940) 111.

[6] Araki, H., *Einführung in die Axiomatische Quantenfeldtheorie I, II* (ETH-Lecture, ETH, Zurich, 1962); *Prog. Theor. Phys.* **64** (1980) 719 (this paper is closely related to the many-Hilbert-space theory of measurement); *Mathematics of Quantum Fields* (Iwanami, Tokyo, 1993) [in Japanese].

[7] Araki, H., Y. Munakata, M. Kawaguchi and T. Goto, *Prog. Theor. Phys.* **17** (1957) 419.

[8] Araki, H. and M.M. Yanase, *Phys. Rev.* **120** (1960) 622.

[9] Aspect, A., P. Dalibard and G. Roger, *Phys. Rev. Lett.* **49** (1982) 1804.

[10] Audretsch, J., M. Mensky and V. Namiot, *Phys. Lett.* **A203** (1995) 267.

[11] Badurek, G., H. Rauch and J. Summhammer, *Phys. Rev. Lett.* **51** (1983) 1015; G. Badurek, H. Rauch and D. Tuppinger, *Phys. Rev.* **34A** (1986) 2600.

[12] Badurek, G., H. Rauch and A. Zeilinger, *Proc. Workshop Neutron Spin Echo*, Grenoble, 1979 [*Lect. Notes Phys.* **128** (1980) 136 (Springer-Verlag, Berlin, 1979)].

[13] Badurek, G., H. Rauch and A. Zeilinger, *Matter Wave Interferometry* (North-Holland, Amsterdam, 1988).

[14] Ballentine, L. E., *Phys. Rev.* **A43** (1991) 5165.

[15] Beige, A. and G. Hegerfeldt, *Phys. Rev.* **A53** (1996) 53.

[16] Belavkin, V.P., *Phys. Lett.* **A140** (1989) 355; *J. Phys.* **A22** (1989) L1109.

[17] Bell, J.S., *Physics* **1** (1965) 195.

[18] Bell, J.S., *Helv. Phys. Acta* **48** (1975) 93.

[19] Bernardini, C., L. Maiani and M. Testa, *Phys. Rev. Lett.* **71** (1993) 2687.

[20] Berry, M., in *Fundamental Problems in Quantum Theory*, eds. D.M. Greenberger and A. Zeilinger (*Ann. N.Y. Acad. Sci.* **755**, New York, 1995), p.303.

[21] Beskow, A. and J. Nilsson, *Arkiv für Fysik* **34** (1967) 561.

[22] Blanchard, Ph. and A. Jadczyk, *Phys. Lett.* **A183** (1993) 272.

[23] Blasi, R., H. Nakazato, M. Namiki and S. Pascazio, *Phys. Lett.* **A223** (1996) 320; "Emergence of a Wiener process as a result of the quantum mechanical interaction with a macroscopic medium," to appear in *Physica*.

[24] Bogoliubov, N.N., V.V. Tolmachev and D.V. Shirkov, *A New Method in the Theory of Superconductivity* (Academy of Sciences of U.S.S.R., Moscow, 1958).

[25] Bohm, D., *Quantum Theory* (Prentice-Hall, Englewood Cliffs, N.J., 1951), p.614.

[26] Bohm, D., *Phys. Rev.* **85** (1952) 166, 180; D. Bohm and J.P. Vigier, *Phys. Rev.* **96** (1954) 208; **109** (1958) 1882.

[27] Bohr, N., *Atti del Congresso Internazionale dei Fisici* (Zanichelli, Bologna, 1928) vol.2, p.565; *Nature* **121** (1928) 580; *Atomic Theory and the Description of Nature* (Cambridge University Press, 1934, 1961), p.52. See also *Collected Works*, ed. R. Peierls (North-Holland, Amsterdam, 1986).

[28] Bohr, N., *Phys. Rev.* **48** (1935) 696.

[29] Bonse, U. and W. Graeff, in *Topics in Applied Physics* **22**, ed. by H.-J. Queisser (Springer, Berlin, 1977), p.93.

[30] Bonse, U. and H. Rauch, eds., *Neutron Interferometry* (Clarendon, Oxford, 1979).

[31] Braginsky, V.B. and Y.I. Volontsov, *Sov. Phys. Usp.* **17** (1975) 644.

[32] Braginsky, V.B. and F.Y. Khalili, in *Quantum Measurement*, ed. K.S. Thorne (Cambridge Univ. Press, Cambridge, 1992).

[33] Breit, G. and E.P. Wigner, *Phys. Rev.* **49** (1936) 519.

[34] Busch, P., P.J. Lahti and P. Mittelstaedt, *The quantum theory of measurement* (Springer-Verlag, Berlin, 1991).

[35] Chiu, C.B., B. Misra and E.C.G. Sudarshan, *Phys. Lett.* **117B** (1982) 34.

[36] Cho, G-C., H. Kasari and Y. Yamaguchi, *Prog. Theor. Phys.* **90** (1993) 803.

[37] Cini, M., *Nuovo Cim.* **73B** (1983) 27.

[38] Clauser, J.F., *Phys. Rev. Lett.* **37** (1976) 1223.

[39] Clauser, J.F. and A. Shimony, *Rep. Prog. Phys.* **41** (1978) 1881.

[40] Clothier, R., H. Kaiser, S.A. Werner, H. Rauch and H. Wölwitsch, *Phys. Rev.* **A44** (1991) 5357.

[41] Colella, W., A.W. Overhauser and S.A. Werner, *Phys. Rev. Lett.* **34** (1975) 1472.

[42] Comsa, G., *Phys. Rev. Lett.* **51** (1983) 1105.

[43] Cook, R.J., *Phys. Scr.* **T21** (1988) 49.

[44] Costa de Beauregard, O., *Found. Phys.* **22** (1992) 121.

[45] Daneri, A., A. Loinger and G.M. Prosperi, *Nucl. Phys.* **33** (1962) 297.

[46] DeGasperis, A, L. Fonda and G.C. Ghirardi, *Nuovo Cim.* **A21** (1974) 471.

[47] d'Espagnat, B., *Conceptual Foundations of Quantum Mechanics* (Benjamin, New York, 1971).

[48] Dewdney, C., P. Gueret, A. Kyprianidis and J.P. Vigier, *Phys. Lett.* **102A** (1984) 291.

[49] Dewdney, C. and P.R. Holland, in *Quantum mechanics versus local realism*, ed. by F. Selleri (Plenum, New York, 1988), p.301.

[50] Diósi, L., *Phys. Lett.* **A129** (1988) 419; **A132** (1988) 233; *Phys. Rev.* **A40** (1988) 1165.

[51] Drabkin, G.M. and R.A. Zhitnikov, *Sov. Phys. JETP* **11** (1960) 729.

[52] Ebisawa, T., S. Tasaki, T. Kawai, M. Hino, T. Akiyoshi, N. Achiwa, Y. Otake and H. Funahashi, in *Proc. Int. Symp. on Advance in Neutron Optics and Related Research Facilities*, eds. S. Kawano, T. Kawai and A. Kawaguchi (Phys. Soc. Japan, Tokyo, 1996), p.66 [*J. Phys. Soc. Jpn. (Suppl. A)* **65** (1996) 66].

[53] Einstein, A., B. Podolsky and N. Rosen, *Phys. Rev.* **47** (1935) 777.

[54] Everett III, H., *Rev. Mod. Phys.* **29** (1957) 454. See also, *The Many-Worlds Interpretation of Quantum Mechanics*, eds. B.S. DeWitt and N. Graham (Princeton University Press, Princeton, 1973).

[55] Exner, P., *Open Quantum Systems and Feynman Integrals* (Reidel, Dordrecht, 1985), Chapter I (the notes to Section 1.5 are particularly interesting).

[56] Fermi, E., *Nuclear Physics* (Univ. Chicago, Chicago, 1950) pp. 136, 148; See also *Notes on Quantum Mechanics. A Course Given at the University of Chicago in 1954*, ed. E. Segré (Univ. Chicago, Chicago, 1960) Lec. 23; also *Rev. Mod. Phys.* **4** (1932) 87.

[57] Fock, V. and N. Krylov, *J. Phys.* **11** (1947) 112.

[58] Fonda, L., G.C. Ghirardi and A. Rimini, *Rep. Prog. Phys.* **41** (1978) 587.

[59] Fonda, L., G.C. Ghirardi, A. Rimini and T. Weber, *Nuovo Cim.* **A15** (1973) 689; **A18** (1973) 805.

[60] Fonda, L., G.C. Ghirardi and T. Weber, *Phys. Lett.* **131B** (1983) 309.

[61] Frerichs, V. and A. Schenzle, in *Foundations of Quantum Mechanics*, eds. T.D. Black, M.M. Nieto, H.S. Pilloff, M.O. Scully and R.M. Sinclair (World Scientific, Singapore, 1992).

[62] Fry, E.S. and R.C. Thompson, *Phys. Rev. Lett.* **37** (1976) 465.

[63] Furry, W.H., *Boulder Lectures in Theoretical Physics* Vol.8A (University Colorado Press, 1966).

[64] Gamow, G., *Z. Phys.* **51** (1928) 204.

[65] Gaveau, B. and L.S. Schulman, *J. Stat. Phys.* **58** (1990) 1209; *J. Phys.* **A28** (1995) 7359.

[66] Ghirardi, G.C., C. Omero, T. Weber and A. Rimini, *Nuovo Cim.* **A52** (1979) 421.

[67] Ghirardi, G.C., A. Rimini and T. Weber, *Phys. Rev.* **D34** (1986) 470; **D36** (1987) 3287; G.C. Ghirardi, P. Pearle and A. Rimini, *Phys. Rev.* **A42** (1990) 78; G.C. Ghirardi and A. Rimini, in *Sixty Two Years of Uncertainty*, ed. A. Miller (Plenum, New York, 1990).

[68] Goldberger, M.L. and F. Seitz, *Phys. Rev.* **71** (1974) 294.

[69] Green, H.S., *Nuovo Cim.* **9** (1958) 880.

[70] Greenberger, D.M. and A.W. Overhauser, *Rev. Mod. Phys.* **51** (1979) 43; D.M. Greenberger, *Rev. Mod. Phys.* **55** (1983) 875.

[71] Haroche, S., *Nuovo Cim.* **110B** (1995) 501; in *Proc. International Conference "Fundamental Problems in Quantum Theory"* (*Ann. NY Acad. Sci.* **755**, 1995), p.73.

[72] Hayakawa, S., T. Sasakawa, K. Tomita, M. Yasuno and M. Yokota, *Prog. Theor. Phys.* **21** (1959) 85.

[73] Heinrich, M., H. Rauch and H. Wolwitsch, *Physica* **B156-157** (1989) 588.

[74] Heisenberg, W., *Z. Phys.* **43** (1927) 172.

[75] Hellmuth, T., H. Walther, A. Zajonic and W. Schleich, *Phys. Rev.* **A35** (1987) 2532; in *Proc. Second Int. Symp. Foundations of Quantum Mechanics*, eds. M. Namiki *et al.* (Phys. Soc. Japan, Tokyo, 1987), p.25.

[76] Hellund, E.J., *Phys. Rev.* **89** (1953) 919.

[77] Hepp, K., *Helv. Phys. Acta* **45** (1972) 237.

[78] Hiyama, K. and S. Takagi, *Phys. Rev.* **A48** (1993) 2568.

[79] Holland, P.R., *The Quantum Theory of Motion* (Cambridge University Press, Cambridge, 1993).

[80] Home, D. and M.A.B. Whitaker, *J. Phys.* **A25** (1992) 657; *Phys. Lett.* **A173** (1993) 327.

[81] Home, D. and M.A.B. Whitaker, "A Conceptual Analysis of Quantum Zeno: Paradox, Measurement and Experiment," preprint 1997.

[82] Horwitz, L.P. and E. Katznelson, *Phys. Rev. Lett.* **50** (1983) 1184.

[83] Inagaki, S., M. Namiki and T. Tajiri, *Phys. Lett.* **A166** (1992) 5.

[84] Itano, W.M., D.J. Heinzen, J.J. Bollinger and D.J. Wineland, *Phys. Rev.* **A41** (1990) 2295.

[85] Itano, W.M., D.J. Heinzen, J.J. Bollinger and D.J. Wineland, *Phys. Rev.* **A43** (1991) 5168.

[86] Itano, W.M., *et al.*, in *Proc. Int. Symp. Quantum Physics and the Universe*, eds. M. Namiki *et al.* (Pergamon, Oxford, 1993), p.169.

[87] Jacobson, D.L., S.A. Werner and H. Rauch, *Phys. Rev.* **A49** (1994) 3196.

[88] Jammer, M., *Conceptional Development of Quantum Mechanics* (McGraw-Hill, New York, 1966).

[89] Jauch, J.M., *Helv. Phys. Acta* **37** (1964) 293.

[90] Jauch, J.M., E.P. Wigner and Y.Y. Yanase, *Nuovo Cim.* **48B** (1967) 144.

[91] Joos, E. and H.D. Zeh, *Z. Phys.* **B59** (1984) 223.

[92] Kabir, P. K., *The CP Puzzle: Strange Decays of the Neutral Kaon*, (Academic Press, New York, 1968) (Appendix on Wigner-Weisskopf method).

[93] Kaiser, H., R. Clothier, S.A. Werner, H. Rauch and H. Wölwitsch, *Phys. Rev.* **A45** (1992) 31.

[94] Kaiser, H., S.A. Werner and E.A. George, *Phys. Rev. Lett.* **50** (1983) 560.

[95] Kakazu, K. and S. Matsumoto, *Phys. Rev.* **A42** (1990) 5093.

[96] Kamei, O., T. Kobayashi and M. Namiki, *Prog. Theor. Phys.* **28** (1962) 233.

[97] Kawai, T., T. Ebisawa, S. Tasaki, T. Akiyoshi, M. Hino, A. Achiwa, Y. Otake and H. Funahashi, in *Proc. Int. Symp. on Advance in Neutron Optics and Related Research Facilities*, eds. S. Kawano, T. Kawai and A. Kawaguchi (Phys. Soc. Japan, Tokyo, 1996), p.230 [*J. Phys. Soc. Jpn. (Suppl. A)* **65** (1996) 230].

[98] Kendrick, H., J.S. King, S.A. Werner and A. Arrot, *Nucl. Instr. Meth.* **79** (1970) 82.

[99] Khalfin, L.A., *Dokl. Acad. Nauk USSR* **115** (1957) 277 [*Sov. Phys. Dokl.* **2** (1957) 340]; *Zh. Eksp. Teor. Fiz.* **33** (1958) 1371 [*Sov. Phys. JETP* **6** (1958) 1053].

[100] Khalfin, L.A., *Zh. Eksp. Teor. Fiz. Pis. Red.* **8** (1968) 106 [*JETP Letters* **8** (1968) 65]; *Usp. Fiz. Nauk.* **160** (1990) 185 [*Sov. Phys. Usp.* **33** (1990) 10].

[101] Khalfin, L.A., *Phys. Lett.* **112B** (1982) 223.

[102] Kiang, D., *Am. J. Phys.* **42** (1974) 785.

[103] Kiess, T.E., Y.H. Shih, A.V. Sergienko and C.O. Alley, *Phys. Rev. Lett.* **71** (1993) 3893.

[104] Kittel, C., *Introduction to Solid State Physics* (John Wiley & Sons, New York, 1966).

[105] Klein, A.G., G.I. Opat and W.A. Hamilton, *Phys. Rev. Lett.* **50** (1983) 563.

[106] Klein, A.G. and S.A. Werner, *Rep. Prog. Phys.* **46** (1983) 259.

[107] Kono, N., K. Machida, M. Namiki and S. Pascazio, *Phys. Rev.* **A54** (1996) 1064.

[108] Kowalski, J.M. and J.L. Fry, *J. Math. Phys.* **28** (1987) 2407.

[109] Kudaka, S., S. Matsumoto and K. Kakazu, *Prog. Theor. Phys.* **82** (1989) 665.

[110] Kwiat, P., H. Weinfurter, T. Herzog, A. Zeilinger and M. Kasevich, in *Fundamental Problems in Quantum Theory*, eds. D.M. Greenberger and A. Zeilinger (*Ann. N.Y. Acad. Sci.* **755**, New York, 1995), p.383; *Phys. Rev. Lett.* **74** (1995) 4763.

[111] Lepore, V.L., *Phys. Rev.* **A50** (1994) 5014.

[112] Li, Y.Q. and Y.X. Chen, *Phys. Lett.* **A202** (1995) 325.

[113] Lichtenberg, A.J. and M.A. Lieberman, *Regular and Chaotic Dynamics* (Appl. Math. Series **38**, Springer, 1983, 1992).

[114] Liu, X.J. and C.P. Sun, *Phys. Lett.* **A198** (1995) 371.

[115] Loinger, A., *Nucl. Phys.* **A108** (1968) 245.

[116] Loudon, R., *Quantum Theory of Light*, 2nd. ed. (Oxford University Press, Oxford, 1983).

[117] Machida, S. and M. Namiki, *Prog. Theor. Phys.* **63** (1980) 1457, 1833.

[118] Machida, S. and M. Namiki, in *Proc. Int. Symp. on Foundations of Quantum Mechanics*, eds. S. Kamefuchi *et al.* (Phys. Soc. Japan, Tokyo, 1984), p.136; in *Selected Papers from the Proceedings of the First through Fourth International Symposia on Foundations of Quantum Mechanics in the Light of New Technology*, eds. S. Nakajima, Y. Murayama and A. Tonomura (World Scientific, Singapore, 1996).

[119] Mandel, L. and E. Wolf, *Optical Coherence and Quantum Optics* (Cambridge Univ. Press, Cambridge, 1995).

[120] Mandelstam, L. and I. Tamm, *J. Phys.* **9** (1945) 249.

[121] McVoy, K.M., *Ann. Phys.* (New York) **54** (1969) 552.

[122] Merzbaker, E., *Quantum Mechanics* (John Wiley & Sons, New York, 1970).

[123] Messiah, A., *Mécanique Quantique I, II* (Dunod, Paris, 1959) [*Quantum Mechanics I, II* (North-Holland, Amsterdam, 1961)].

[124] Mihokova, E., S. Pascazio and L.S. Schulman, "Hindered decay: Quantum Zeno effect through electromagnetic field domination," *Phys. Rev. A*, in print.

[125] Misra, B. and E.C.G. Sudarshan, *J. Math. Phys.* **18** (1977) 756.

[126] Monroe, C., D.M. Meekhof, B.E. King and D.J. Wineland, *Science* **272** (1996) 1131.

[127] Murayama, Y., *Found. Phys. Lett.* **3** (1990) 103; Y. Murayama and M. Namiki, in *The concept of probability*, E.I. Bitsakis and C.A. Nicolaides eds. (Kluwer Academic, Dordrecht, 1989).

[128] Nabekawa, Y., M. Namiki and S. Pascazio, *Phys. Lett.* **167A** (1992) 435.

[129] Nakazato, H., K. Machida, M. Namiki and S. Pascazio, in *Proc. Int. Symp. on Advance in Neutron Optics and Related Research Facilities*, eds. S. Kawano, T. Kawai and A. Kawaguchi (Phys. Soc. Japan, Tokyo, 1996), p.33 [*J. Phys. Soc. Jpn. (Suppl. A)* **65** (1996) 33].

[130] Nakazato, H., M. Namiki and S. Pascazio, *Phys. Rev. Lett.* **73**, 1063 (1994).

[131] Nakazato, H., M. Namiki, S. Pascazio and H. Rauch, *Phys. Lett.* **A199** (1995) 27.

[132] Nakazato, H., M. Namiki and S. Pascazio, *Int. J. Mod. Phys.* **B10** (1996) 247.

[133] Nakazato, H., M. Namiki, S. Pascazio and H. Rauch, *Phys. Lett.* **A217** (1996) 203.

[134] Nakazato, H., M. Namiki, S. Pascazio and Y. Yamanaka, *Phys. Lett.* **A222** (1996) 130.

[135] Nakazato, H. and S. Pascazio, *Phys. Lett.* **A156** (1991) 386; *Phys. Rev.* **A45** (1992) 4355.

[136] Nakazato, H. and S. Pascazio, *Phys. Rev. Lett.* **70** (1993) 1; *Phys. Rev.* **A48** (1993) 1066.

[137] Nakazato, H. and S. Pascazio, *Phys. Lett.* **A192** (1994) 169.

[138] Namiki, M., *Found. Phys.* **18** (1988) 29; in *Microphysical Reality and Quantum Mechanics*, eds. F. Selleri, G. Tarrozi and A. van der Merwe (Reidel, Dordrecht, 1987), p.3; in *Quantum Mechanics of Macroscopic Systems and Measurement Problem*, ed. Y. Otsuki (Kyoritu, Tokyo, 1985), p.139 [in Japanese]; *Ann. NY Acad. Sci.* **480** (1986) 78.

[139] Namiki, M. and N. Mugibayashi, *Prog. Theor. Phys.* **10** (1953) 474.

[140] Namiki, M., Y. Otake and H. Soshi, *Prog. Theor. Phys.* **77** (1987) 508.

[141] Namiki, M. and S. Pascazio, *Phys. Lett.* **147A** (1990) 430.

[142] Namiki, M. and S. Pascazio, *Phys. Rev.* **A44** (1991) 39.

[143] Namiki, M. and S. Pascazio, *Found. Phys.* **22** (1992) 451.

[144] Namiki, M. and S. Pascazio, *Phys. Rep.* **232** (1993) 301.

[145] Namiki, M., S. Pascazio, and H. Rauch, *Phys. Lett.* **173A** (1993) 87.

[146] Namiki, M., S. Pascazio and C. Schiller, *Phys. Lett.* **A182** (1994) 17.

[147] Ne'eman, Y., in *Microphysical Reality and Quantum Formalism*, ed. A. van der Merwe *et al.* (Kluwer, 1988), p.145; in *Proc. of the Symposium on the Foundation of Quantum Mechanics*, eds. P. Busch *et al.* (World Scientific, Singapore, 1993), p.289.

[148] Onofrio, R., C. Presilla and U. Tambini, *Phys. Lett.* **A183** (1993) 135.

[149] Ou, Z.Y. and L. Mandel, *Phys. Rev. Lett.* **61** (1988) 50; Z.Y. Ou, C.K. Hong and L. Mandel, *Opt. Comm.* **67** (1988) 159.

[150] Overhauser, A.W. and W. Colella, *Phys. Rev. Lett.* **33** (1974) 1237.

[151] Paley, R.E.A.C. and N. Wiener, *Fourier Transforms in the Complex Domain* (American Mathematical Society Colloquium Publications Vol. XIX, New York, 1934).

[152] Pascazio, S. and M. Namiki, *Phys. Rev.* **A50** (1994) 4582.

[153] Pascazio, S., M. Namiki, G. Badurek and H. Rauch, *Phys. Lett.* **A179** (1993) 155.

[154] Pauli, W., *Festschrift zum 60. Geburtstage A. Sommerfelds* (Hirzel, Leipzig, 1928), p.30.

[155] Pati, A.K., *Phys. Lett.* **A215** (1996) 7.

[156] Pearle, P., *Phys. Rev.* **A39** (1989) 2277.

[157] Peres, A., *Am. J. Phys.* **48** (1980) 931; *Ann. Phys.* **129** (1980) 33.

[158] Peres, A. and A. Ron, *Phys. Rev.* **A42** (1990) 5720.

[159] Perrie, W., A.J. Duncan, H.J. Beyer and H. Kleinpoppen, *Phys. Rev. Lett.* **54** (1985) 1790.

[160] Petrosky, T., S. Tasaki and I. Prigogine, *Phys. Lett.* **A151** (1990) 109; *Physica* **A170** (1991) 306.

[161] Rauch, H., in *Proc. Int. Symp. on Foundations of Quantum Mechanics*, eds. S. Kamefuchi *et al.* (Phys. Soc. Japan, Tokyo, 1984), p.277; in *Proc. Second Int. Symp. on Foundations of Quantum Mechanics*, eds. M. Namiki *et al.* (Phys. Soc. Japan, Tokyo, 1987), p.3; in *Proc. Third Int. Symp. on Foundations of Quantum Mechanics*, eds. S. Kobayashi *et al.* (Phys. Soc. Japan, Tokyo, 1990), p.3; in *Selected Papers From the Proceedings of the First through Fourth International Symposia on Foundations of Quantum Mechanics in the Light of New Technology*, eds. S. Nakajima, Y. Murayama and A. Tonomura (World Scientific, Singapore, 1996).

[162] Rauch, H., *Nuovo Cim.* **B110** (1995) 557.

[163] Rauch, H. and D. Petrascheck, in *Topics in Current Physics* 6 (Springer, Berlin, 1979) Chapter 9.

[164] Rauch, H., J. Summhammer, M. Zawisky and E. Jericha, *Phys. Rev.* **A42** (1990) 3726.

[165] Rauch, H., W. Treiner and U. Bonse, *Phys. Lett.* **A47** (1974) 369.

[166] Rauch, H, H. Wölwitsch, H. Kaiser, R. Clothier and S.A. Werner, *Phys. Rev.* **A53** (1996) 902.

[167] Rempe, G., W. Schleich, M.O. Scully and H. Walther, in *Proc. Third Int. Symp. Foundations Quantum Mechanics*, eds. S. Kobayashi *et al.* (Phys. Soc. Japan, Tokoy, 1990), p.294; in *Selected Papers From the Proceedings of the First through Fourth International Symposia on Foundations of Quantum Mechanics in the Light of New Technology*, eds. S. Nakajima, Y. Murayama and A. Tonomura (World Scientific, Singapore, 1996).

[168] Renninger, M., *Zeit. Phys.* **158** (1960) 417.

[169] Rosenfeld, L., *Prog. Theor. Phys. Suppl. Extra Number* (1965) 222.

[170] Rosenfeld, L., *Nucl. Phys.* **A108** (1968) 241.

[171] Rozman, M.G., P. Reineker and R. Tehver, *Phys. Rev.* **A49** (1994) 3310.

[172] Russell, B., *A History of Western Philosophy* (George Allen & Unwin, London, 1946).

[173] Saito, N., in *Quantum Physics, Chaos Theory, and Cosmology*, eds. M. Namiki *et al.* (AIP, New York, 1996), p.275.

[174] Schärpf, O., *Physica* **B156-157** (1989) 631.

[175] Schrödinger, E., *Proc. Cambridge Phil. Soc.* **31** (1935) 555.

[176] Schulman, L.S., A. Ranfagni and D. Mugnai, *Phys. Scr.* **49** (1994) 536.

[177] Scully, M.O. and H. Walther, *Phys. Rev.* **A39** (1989) 5229.

[178] Selleri, F. and G. Tarozzi, *Riv. Nuovo Cim.* **4** (1981) 35; *Quantum Mechanics versus Local Realism: The Einstein, Podolsky and Rosen paradox*, F. Selleri ed. (Plenum Press, New York, 1988).

[179] Stapp, H.P., *Nuovo Cim.* **B40** (1977) 191.

[180] Staudenmann, J.-L., S.A. Werner, R. Colella and A.W. Overhauser, *Phys. Rev.* **A21** (1980) 1419.

[181] Stern, A., Y. Aharonov and Y. Imry, *Phys. Rev.* **A41** (1990) 3436.

[182] Stern, A., Y. Aharonov, A. Yacoby and Y. Imry, in *Proc. Third Int. Symp. Foundations of Quantum Mechanics*, eds. S. Kobayashi *et al.* (Phys. Soc. Japan, Tokyo, 1990), p.201; in *Selected Papers From the Proceedings of the First through Fourth International Symposia on Foundations of Quantum Mechanics in the Light of New Technology*, S. Nakajima, Y. Murayama and A. Tonomura eds. (World Scientific, Singapore, 1996).

[183] Stratonovich, R.L. and V.P. Belavkin, *Int. J. Theor. Phys.* **35** (1996) 2215; R.L. Stratonovich, "On the dynamical interpretation for the collapse of state during quantum measurements," Mathematics Preprint Series, Univ. Nottingham, 1996.

[184] Summhammer, J., G. Badurek, H. Rauch and U. Klischko, *Phys. Lett.* **A90** (1982) 110; J. Summhammer, G. Badurek, H. Rauch, U. Klischko and A. Zeilinger, *Phys. Rev.* **A27** (1983) 2523.

[185] Summhammer, J., H. Rauch and D. Tuppinger, *Phys. Rev.* **A36** (1987) 4447.

[186] Sun, C.P., *Phys. Rev.* **A48** (1993) 898.

[187] Takabayashi, T., in *Proc. Int. Symp. on the Foundations of Quantum Mechanics*, eds. S. Kamefuchi *et al.* (Phys. Soc. Japan, Tokyo 1984), p.44.

[188] Taubes, G., *Science* **274** (1996) 1615.

[189] Tomita, K., *J. Phys. Soc. Jap. Suppl. B-I* **17** (1962) 71.

[190] Tonomura, A., J. Endo, T. Matsuda and T. Kaesaki, *Am. J. Phys.* **57** (1989) 117.

[191] Van Hove, L., *Physica* **21** (1955) 517.

[192] Venugopalan, A., *Phys. Rev.* **A50** (1994) 2742; A. Venugopalan, D. Kumar and R. Ghosh, *Physica* **A220** (1995) 563, 576; *Current Science* **68** (1995) 62.

[193] Venugopalan, A. and R. Ghosh, *Phys. Lett.* **A204** (1995) 11.

[194] Vigier, J.P., *Pramana J. Phys.* **25** (1985) 397; J.P. Vigier and S. Roy, *Hadronic J. Suppl.* **1** (1985) 474.

[195] von Neumann, J., *Die Mathematische Grundlagen der Quantenmechanik* (Springer, Berlin, 1932) [English translation: *Mathematical Foundations of Quantum Mechanics*, translated by E.T. Beyer (Princeton University Press, Princeton, 1955)].

[196] von Neumann, J., Ref. 195, p.195 [p.366 of the English translation].

[197] Walls, D.F. and G.J. Milburn, *Quantum Optics* (Springer, Berlin, 1994).

[198] Watanabe, S., private communication.

[199] Webb, R.A., S. Washburn, A.D. Benoit, C.P. Umbach and R.B. Laibowitz, in *Proc. 2nd Int. Symp. Foundations of Quantum Mechanics*, eds. M. Namiki *et al.* (Phys. Soc. Japan, Tokyo, 1986), p.193.

[200] Weisskopf, V. and E.P. Wigner, *Z. Phys.* **63** (1930) 54; **65** (1930) 18.

[201] Werner, S.A., *Phys. Today* **33** (1980) 24.

[202] Wheeler, J.A., *Rev. Mod. Phys.* **29** (1957) 463.

[203] Wheeler, J.A., in *Mathematical Foundations of Quantum Theory*, ed. R. Marlow (Academic Press, New York, 1978), p.9; in *Proc. Int. Symp. on the Foundations of Quantum Mechanics*, eds. S. Kamefuchi *et al.* (Phys. Soc. Japan, Tokyo, 1984), p.140.

[204] Wheeler, J.A. and W.H. Zurek, eds., *Quantum Theory and Measurement* (Princeton University Press, New Jersey, 1983).

[205] Wigner, E.P., *Am. J. Phys.* **31** (1963) 6.

[206] Wigner, E.P., in *Rendiconti Scuola Internazionale di Fisica "E. Fermi,"* *IL Corso* (Academic Press, New York, 1971), p.1.

[207] Wootters, W.K. and W.H. Zurek, *Phys. Rev.* **19D** (1979) 473.

[208] Yaffe, L.G., *Rev. Mod. Phys.* **54** (1982) 407.

[209] Yanase, M.M., *Phys. Rev.* **123** (1961) 666.

[210] Zeh, H.D., in *Enrico Fermi School of Physics IL*, ed. B. d'Espagnat (Academic Press, New York, 1971), p.263.

[211] Zurek, W.H., *Phys. Rev.* **24D** (1981) 1516; **26D** (1982) 1862; *Phys. Today*, **44** (1991) 36.

Index

www.ingramcontent.com/pod-product-compliance
Lightning Source LLC
Chambersburg PA
CBHW050639190326
41458CB00008B/2334